THE HOME
BLACKSMITH

Ryan Ridgway, DVM

DEDICATION

To my daughter Annika: You spent many hours in the forge for this book, with mommy and I, before you were born. Love you forever, my little apprentice.

The Home Blacksmith

CompanionHouse Books™ is an imprint of Fox Chapel Publishers International Ltd.

Project Team
Vice President–Content: Christopher Reggio
Editor: Amy Deputato
Copy Editor: Joann Woy
Design: Mary Ann Kahn
Index: Elizabeth Walker

ISBN 978-1-62008-213-3

Library of Congress Cataloging-in-Publication Data
Names: Ridgway, Ryan, author.
Title: The home blacksmith : tools, techniques, and 40 practical projects for
  the blacksmith hobbyist / Ryan Ridgway, DVM.
Description: Irvine, CA : Lumina Press, [2016] | Includes index.
Identifiers: LCCN 2016017129 | ISBN 9781620082133 (paperback)
Subjects: LCSH: Blacksmithing--Amateurs' manuals. | BISAC: CRAFTS & HOBBIES /
  Metal Work.
Classification: LCC TT221 .R53 2016 | DDC 682--dc23 LC record available at
https://lccn.loc.gov/2016017129

This book has been published with the intent to provide accurate and authoritative information in regard to the subject matter within. While every precaution has been taken in the preparation of this book, the author and publisher expressly disclaim any responsibility for any errors, omissions, or adverse effects arising from the use or application of the information contained herein. The techniques and suggestions are used at the reader's discretion.

Fox Chapel Publishing
903 Square Street
Mount Joy, PA 17552

Fox Chapel Publishers International Ltd.
7 Danefield Road, Selsey (Chichester)
West Sussex PO20 9DA, U.K.

www.facebook.com/companionhousebooks

We are always looking for talented authors. To submit an idea, please send a brief inquiry to acquisitions@foxchapelpublishing.com.

Printed and bound in China
21 20 19 18   4 6 8 10 9 7 5 3

CONTENTS

INTRODUCTION

I have had the luck of entering blacksmithing when the craft was resurging and the Internet was offering forums where blacksmiths across the globe could congregate around the virtual forge and share tips and teachings. Traditional blacksmiths didn't have that opportunity, relying instead on local guilds and apprenticeships to pass on the trade. In the early 1900s, blacksmithing all but died out as a craft because the knowledge wasn't being shared.

It is said that nothing is ever new in blacksmithing—only rediscovered. The collective of blacksmiths through the ages must have tried everything we know today at least once or twice at some point in history. After all, modern smiths are using these same techniques to create similar objects that all smiths through time have made. Being a part of that tradition is humbling, and I feel grateful to be a part of passing on the knowledge I have accumulated from books and from other smiths, along with tricks I've picked up in my years of blacksmithing on the farm and at museum demonstrations. All of the smiths I have come across both online and in person have helped me become a better blacksmith. I hope that this book helps get you hooked on blacksmithing or, if you are already hooked, gives you some tips that you might not have thought of.

GLOSSARY

ALLOY. A mixture of different metals and molecules.

ANNEAL. To slowly cool hardened steel in a controlled manner to fully relax the grain structure, softening the steel.

ANVIL. A heavy block of steel for hammering on.

APPRENTICE. A student taken on by a master to learn the trade by experience. Historically, while learning, apprentices did the more mundane tasks, such as cleaning, often working for free in trade. Once ready, they became journeymen and were able to be employed by different master smiths before attempting to become tested as a master smith.

BLACKSMITH. A smith who forges iron. A whitesmith uses files and scrapers to finish the project rather than leaving it as forged.

BRAZING ROD. A brass or bronze rod used to fuse steel together at high heats; found in most welding shops.

BORAX. Sodium tetraborate, commonly found in laundry aisles but also useful as a flux to prevent oxidation from forming during forge welding.

BOSS. The hinge section of a pair of tongs. Also a raised area on a piece of iron.

BUTCHER. A tool with a shallow one-sided bevel, used to start a sharp shoulder.

CARBON. A molecule that turns iron into steel. The amount of carbon in steel determines how hardenable the steel is. Too much carbon can make steel brittle, as in cast iron.

CASE HARDENING. Adding carbon to the surface of mild steel to make a hardenable skin. There are commercial case-hardening products available, but traditionally it was done by putting the steel in a container of hair, horn, or bone and heating it to red hot for multiple days, depending on how much carbon was wanted in the steel.

CENTER PUNCH. A punch with a sharp point on the tip to create a mark, either for bending or making it easier for a drill bit to start without skating away.

CHAMFER. To flatten or round a corner by forging or filing.

CHARCOAL. Partially burnt wood that has had everything but the carbon burnt off. Used for centuries as the primary forge fuel until coal was discovered.

CHASE. To use punches and chisels to create designs in metal.

CHEEK. The side of the handle hole in a tool.

COLD SHUT. A folded-over piece of metal that creates a crack.

COAL. An organic mineral that can burn. Often has many impurities that melt into slag and clinker but is denser than charcoal, thus requiring less fuel to burn as hot. Two types of blacksmithing coal are used: bituminous (soft) and anthracite (hard). It comes in many different grades depending on its heat production and impurities.

DESCALE. To remove iron oxide created by heat. Often done with a wire brush or by pickling with vinegar or other acids.

DRAW (OUT). To stretch and thin out iron by hammering it; it is generally a uniform shape change rather than tapering.

DRIFT. (v.) To open a punched hole and make it larger or change its shape. (n.) A punch used to open up and refine a drilled or punched hole. Often used with a slot punch or slitting chisel to make larger holes than the width of the steel will allow.

FLUX. A material used to cover the area to be welded and prevent iron oxidation that would inhibit welding. Laundry borax is most often used in forge welding.

FORGE. (v.) To work metal with a hammer. (n.) The area that contains the fire used to heat the metal.

FORGE WELDING. Heating steel to near melting and fusing it with hammer blows.

FULLER. Rounded wedge shape used to control the direction that metal moves while forging. A cross-peen is a good example of a fuller.

HARDEN. To make a piece of steel stronger and more brittle by heating it up and cooling it quickly.

HARDY. An anvil tool used to cut iron, most often hot iron.

HARDY HOLE. A square hole in the anvil used for securing tools, such as the hardy, chisels, and swages; it can also be used to help bend metal or punch holes.

HOT CUT. A chisel for cutting hot iron, either handheld or handled.

HOT RASPING. Using a coarse rasp to quickly remove metal while the piece is hot. It is an efficient method to remove metal, often faster than using a grinding wheel, but it can ruin the hardness of a file.

HUNDREDWEIGHT SYSTEM. An older method of measuring weight; often used on antique anvils.

IRON. An element that makes up the majority of steel alloys. Wrought iron is pure iron with layers of silica in it from being forged from metal extracted from iron ore.

JIG. Any device used to make reproducing a part of a project easier.

LAP WELD. To forge weld pieces of iron together by overlapping them.

LEG VISE. Also known as a post or blacksmith's vise, it is used to hold metal but differs from a machinist's vise by having a leg that takes some of the force off the mounting screws and jaws.

MANDREL. A cone of varying sizes that can be large enough to stand on the floor or can be used in the hardy hole or held by hand. It can be used to make rings or bends. Also known as a beak.

MARTENSITE. A hard form of iron crystal structure in higher carbon steels created by quickly cooling the steel, preventing carbon from precipitating out of the iron crystal structure.

NORMALIZING. To cool steel slowly, but not as slowly as in annealing. It both softens the steel and changes the steel's grain structure less than annealing does.

PEEN (OR PEIN). Rounded end of a hammer, opposite the face. Can be a rounded wedge, as in a cross-peen, or a ball, as in a ball-peen.

PENNY WELD. To flux and use copper to braze two parts together by heating the piece in the forge until the copper melts.

PLANISH. To use light, overlapping hammer blows to smooth and work-harden iron.

PRITCHEL. A slender rectangular punch used by farriers to punch nail holes into horseshoes.

PRITCHEL HOLE. The round hole in the heel of an anvil for punching and drifting.

PUNCH. (v.) To make a hole in a piece of metal as an alternative to drilling. (n.) The tool that punches.

QUENCH. Liquid or gas used to cool steel, most commonly water, oil, and air.

REINS. The handle portion of a pair of tongs.

SCALE. Gray flakes of oxide that form when steel is heated to forging temperatures. It will prevent welds from forming.

SCARF. (v.) To prepare an area for welding by upsetting it and tapering the edges to blend well. (n.) The part of the iron that has been scarfed.

SET DOWN/SETTING DOWN. To forge a step into the stock your are using.

SET HAMMER. A type of hammer that is struck by another hammer to create shoulders and corners that are sharper than can be achieved with a hand hammer.

SHOULDER. A sudden change in dimensions, usually 90 degrees, to create a step or higher point.

SMITHY. A blacksmith's shop.

SPARK TESTING. To rudimentarily determine the carbon content of an unknown steel by grinding it and looking at the sparks. The higher the carbon content, the more of a starburst the sparks will create.

SWAGE. Anvil tool that is used to create different shapes in steel. Can also be a stand-alone block with various shapes cut into it. Most common are V and half-round swages. They can be combined top and bottom to make consistently sized round and square sections.

TEMPER. To control the heating of hardened steel to reduce the amount of brittleness by altering the grain structure slightly.

TEMPER COLORS. Iron oxidation color changes that occur on polished steel when heated to specific temperatures.

TEMPLATE. A pattern used to cut or forge out a repeatable shape.

TUYERE. The air inlet to the forge fire. Connects the air source with the firepot.

UPSET. To increase the dimensions of a piece of steel by forcing it back into itself.

WORK-HARDEN. To hammer the iron cold to harden it slightly. Can lead to brittleness and cracks.

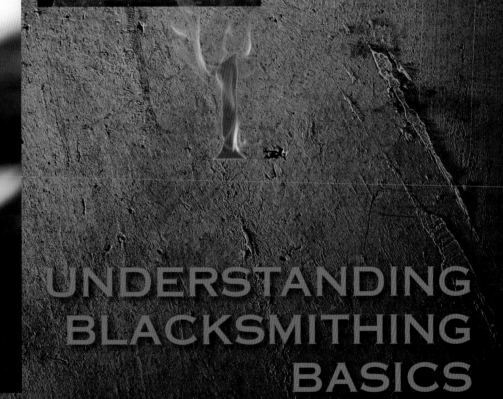

UNDERSTANDING
BLACKSMITHING
BASICS

A BRIEF HISTORY OF BLACKSMITHING

For new blacksmiths, knowing how our craft has evolved is extremely important in helping you set up your first forge without a lot of cost and time spent looking for materials. Ask a number of blacksmiths from around the world what the tools in a blacksmith's forge should look like, and you will get a dozen different answers. This is because, over time, blacksmithing tools have evolved differently in different parts of the world. Ask a North American or European smith what an anvil should look like, and he or she will describe a

stereotypical London-pattern anvil with a horn and a heel, straight from the cartoons. At the same time, an Asian smith may say that it is simply a square block of steel. If you went back to the Middle Ages, a European smith would give you the same answer as the Asian smith. Go back even further and any smith would point to a large, flat rock.

Today, beginner Western blacksmiths get caught up trying to find a London-pattern anvil, thinking that it is the only style of anvil that will work. This often leads to frustration, which can further lead a blacksmith to either purchase an inferior cast-iron anvil from the local welding shop or spend too much money when a simple block of steel would work. Others will spend more time looking for a good piece of railway steel, which has become the ubiquitous makeshift beginner anvil. I have used everything from Peter Wright and Hay-Budden anvils to simple steel cutoffs and hydraulic shafts stood on end. The railway anvils I have used are often as bad as cast iron because they don't have the proper mass underneath the hammering surface to support heavy forging, instead flexing and absorbing too much of your hammer's blow.

I discuss in more detail what to look for in an anvil in a later chapter, but I feel that a little blacksmithing history will help you choose your tools more easily—along with the understanding that it isn't the tools that make you a blacksmith, but rather the techniques. Blacksmiths' tools have changed with the need for and availability of fuel, iron, and science. Your forge can be the same: an evolution based on necessity and availability.

By around 1200 BC, iron was becoming the principal metal for tools and weapons, with some overlapping of the Bronze Age. Prior to this time, iron artifacts were scarce as different cultures experimented with the newly discovered technology. While iron is stronger and makes longer lasting tools, it was slow to become a commonly used metal—and eventually overtake

SAME TECHNIQUES, DIFFERENT TIMES

Realizing that blacksmithing is about the techniques, not the tools, will make setting up your first shop much easier. While the tools may have changed over the years, the techniques that blacksmiths use have remained the same.

bronze—because it was more difficult to produce iron from its rock ore, and it needed higher temperatures to work. Eventually, civilizations advanced their iron-smelting technologies and began using iron for more items, including art. Iron improved a society's production, agricultural, and warfare abilities, and the blacksmith was considered the king of craftsmen because he created the tools for all other trades.

Each area of the world started forging iron with similar tools, likely either rocks or bronze hammers and the same anvils or boulders that they had been using to work bronze and copper. Until relatively recently, in the grand scheme of things, blacksmiths used blocks of steel or stake anvils and hammers that were fairly similar. Around the 1600s, Western European blacksmiths began to develop an anvil shape that most smiths are familiar with: the London or Continental pattern anvil. Unfortunately, this change from centuries past creates many issues for new blacksmiths today because usable antique anvils are hard to find and can be expensive, while good-quality new anvils are even more expensive. All the while, the original anvil shape—a simple block of steel—sits in the scrap yard, waiting to be melted down and recycled.

The forge is another part of the blacksmith's shop that has changed significantly over the past centuries, both in design and in fuel source. An early forge was simply a fire pit fueled by charcoal, and the air supply was likely a pair of leather sacs opened and closed by the blacksmith's assistant, with a wooden or reed pipe connecting the bellows to the fire pit. Over time, this side-draft forge was raised off the ground, and different styles of bellows evolved in different regions.

Eventually, as coal became a more prevalent fuel, bottom-blast forges with blast pipes (tuyeres), known as *duck-nest* tuyeres, became more common and were often sold in mail-order catalogs. Modern blacksmiths now have access to other forge fuels, such as propane and acetylene, as well as electric induction. Later in the book, I will go deeper into the different forge fuels and how to make your own charcoal so that you don't waste time and money trying to find and buy blacksmithing coal.

Historically, blacksmiths forged pure iron, known as *wrought iron*. Wrought iron was created from iron ore with very little added carbon. These days, it is difficult to purchase pure wrought iron, so most blacksmiths use what is known as *mild steel*, a low-carbon steel alloy. Because of this, many smiths use the term "iron" for mild steel because it has replaced pure wrought iron as the most common metal for blacksmithing. This terminology allows for a differentiation between *steel*, which has enough carbon to harden properly for tools, and mild steel, which will not harden. Because you will be making your own tools for the forge and around the house and farm, this is an important distinction.

Now that you are armed with a brief history of what tools and fuels blacksmiths commonly used in the past, we can move on to understanding how metal moves. Once you understand how metal behaves when it's forged, you can find many unconventional tools in scrap yards, just like the blacksmiths of old had to.

THE PHYSICS OF MOVING METAL

Before we get into finding tools and the actual techniques of blacksmithing, it will be helpful to look at how iron behaves when forged, so that we can start to understand how the variously shaped tools move metal. By learning how shapes, rather than specific tools, move metal, your eyes will be opened as you wander through your local scrap yard. In my experience, most beginning blacksmiths end up reading a beginner's blacksmithing book that leads off with a description of the tools in a typical post-1800s blacksmith shop that used coal—and then the new blacksmith spends all of his or her time searching for the tools in the book. Instead, new blacksmiths need to understand how metal can be controlled and then find the tools that have the shapes to do it. This makes setting up a shop much easier.

CONTROLLING METAL'S MOVEMENT

To give you an idea of how metal moves, watch the ripples on water after you throw a round rock into it—the splashes are circular, right? Now splash the water with a stick. The splashes will primarily radiate out to the sides. The same happens with metal. By using differently shaped fullers (wedge-shaped tools), you can control which way the metal is squeezed.

The great thing about iron is its immense strength when cool and its plasticity at higher temperatures—if you heat iron hot enough in a fire, it moves like clay and can be molded easily by a hammer and anvil. Because hot iron moves like clay, a great way to practice without wasting metal is to chill modeling clay in a refrigerator and then shape it into rods of "iron" to forge.

The tools that you use as a blacksmith have two types of shapes: shapes that force metal outward uniformly or shapes that move metal in two directions. Shapes that are uniform in all directions, such as a full-faced hammer or the peen on a ball-peen hammer, push the metal out in all directions. A fuller is any wedge-shaped tool that pushes the metal in two directions.

The thickness of the metal under your hammer affects how far the force of your hammer blow will penetrate. The force of your blow slowly dissipates as it travels through the metal. You will notice this effect when drawing or tapering thicker pieces of iron because the surface moves more than the center, leading to "fish lips," as they are known in blacksmithing circles. To combat this effect, either the metal needs to be hotter, you need to use a heavier hammer, or you need to use a fuller to increase the amount of force per inch.

Blacksmithing is all about controlling how the metal moves so that you can make the most of the short time during which it's hot enough to forge safely. Now that you understand how metal reacts to your tools, you will need to learn another skill to become a proficient blacksmith: hammer control.

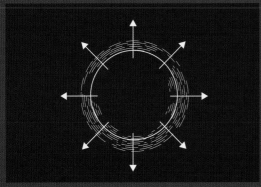

Your hammer face pushes the metal in all directions, as shown, while a fuller pushes the metal in only two directions.

An efficient blacksmith makes the most out of his or her tools. The corners of your hammer and anvil create fullers to move the iron lengthwise rather than spreading it. They can also be used to isolate metal, as shown.

You can use a curved shape, like the horn of an anvil, to make the metal wider than it is long. This is very important when making leaf-shaped elements.

This is an exaggeration of what your hammer face does on the flat of the anvil, spreading the metal inefficiently in all directions. You'd then need to correct either the width or length, depending on the project, with heat and hammer blows.

The force of your blow dissipates as it travels through the metal so that the metal closest to the point of contact spreads faster.

Because the force of your blow dissipates as it travels into the metal, beware of "fish lips" when working thicker pieces of metal because they can become cold shuts once you taper the metal down.

HAMMER CONTROL

You may find it strange to be discussing hammer control before setting up your shop, but I want it to be at the forefront in your mind, along with how metal moves under various shapes. New blacksmiths often get caught up in the need for special tooling and jigs, blaming a lack of jigs when they can't make repetitive pieces, all the while forgetting that they need to practice their hammer control. A blacksmith needs to be efficient with his hammer to forge iron successfully for any length of time.

All blacksmiths are prone to developing repetitive-strain injuries in the hands, elbows, and shoulders of their hammer arms. On top of that, time is money—for blacksmiths, in the form of increased fuel costs. You need to maximize your productivity every time the iron is hot so you won't need to spend time waiting for the iron to reheat.

Forging iron utilizes different hand-eye coordination than most people are used to, and you need to be precise to be efficient and end up with a good-looking project. Time spent practicing your hammer control comes back in spades when you are actually forging. I was fortunate enough to have grown up helping my father in construction, so I made a natural switch from hammering nails to hammering hot iron.

HAMMER CONTROL BEGINS WITH THE HANDLE

To better control your hammer and reduce the amount of strain on your wrist, forearm, and elbow, flatten the sides of your hammer handle. This simple step makes it easier to keep your hammer from twisting in your hand and thus enables you to relax your forearm, preventing tendonitis in your elbow or carpal tunnel syndrome in your wrist. It also lets you easily adjust the angle of your hammer face to use the corner as a fuller.

Traditionally, blacksmith apprentices were given repetitive tasks, such as making nails, to develop their hammer control when they were ready to begin learning to forge. Blacksmiths today can also practice hammering in nails with a hammer (try working up to using the rounded end of a ball-peen hammer) or making nails like they did in days gone by. The important thing is to focus on improving your hammer control, not just making projects. Too often, I see new smiths become frustrated that their projects don't look as nice as those made by experienced smiths, but they won't focus on their hammer control. Of course, it comes with practice, but only if you work at perfecting it; otherwise, you will be doomed to keep making the same mistakes (although you will become very good at those mistakes, I must say!).

PREP FOR SUCCESS

You can minimize your hammer marks by properly preparing the hammer faces and peens. To do this, smooth all sharp edges and radius all corners with a piece of sanding paper or a slack-belted grinder. There should be a slight dome to the face of the hammer, which should be polished to be as smooth as possible. Any sharp edges or dents will transfer from the hammer to your work. The same goes for your anvil.

HAMMER-SWING ERGONOMICS

It is important that you start with good ergonomics to avoid injury as well as bad habits that you will need to correct later. I urge you to watch videos of professional smiths working, because it is impossible to describe good ergonomics properly with still photos. Here are some tips about your hammer swing:

- Don't choke your hammer handle—it's already dead, and by gripping it too tightly your tendons will be taking all of the rebound force. Your grip should be loose, and the hammer should pivot at your thumb and forefinger when it rebounds.

- Use your grip to add extra force as you throw the hammer at the metal. Rather than forcing the hammer into the metal, you can use that same thumb-and-forefinger pivot and squeeze the hammer handle with your other fingers to create a snap in the hammer just before it hits.

- Let the hammer rebound; you aren't trying to force the hammer face through the metal with every hit. Most repetitive-strain injuries are caused by this sudden jolt of energy into your tense tendons, causing them to tear slightly.

- Set the anvil at your knuckle height and don't slouch while forging. This will maximize your hammer swing and let you use all parts of your arm and shoulder while preventing back pain.

- Use all of your arm and shoulder, not just your forearm, to lift the hammer; this will prevent tendonitis in your elbow.

What many smiths don't realize is how many different tool shapes you have with a standard hammer and any anvil, including a block of steel. Once you learn proper hammer control, you won't need to spend time making or using a fuller to move your metal efficiently—you'll just use your hammer and anvil (although a few different fullers and other tools will make your life a lot easier). In fact, I made all of the projects in this book with minimal tooling.

I hope you've gained a basic understanding of how blacksmithing has evolved through the years and how metal responds to different tool shapes. Armed with this knowledge, it will be much easier for you to find blacksmithing tools, regardless of whether you are setting up your first shop or simply adding to your tool collection like every good smith does.

THE MANY FULLERS OF A HAMMER AND ANVIL

Many smiths forget how many fuller options they have with their hammer and anvil and think that they need different fullers and butchers to forge like a professional. While those tools do help, most professionals use their hammer and anvil alone for the bulk of forging. You can use both the near and far side of the anvil and the corners of your hammer (Fig. 1) as fullers to isolate sections of bar and more efficiently draw out your metal. By adjusting the angle at which you hold the iron on the edge of the anvil, you can adjust the amount of taper behind the fuller (Figs. 2 and 3). For a sharper fuller divot, hold the work 90 degrees to the anvil corner.

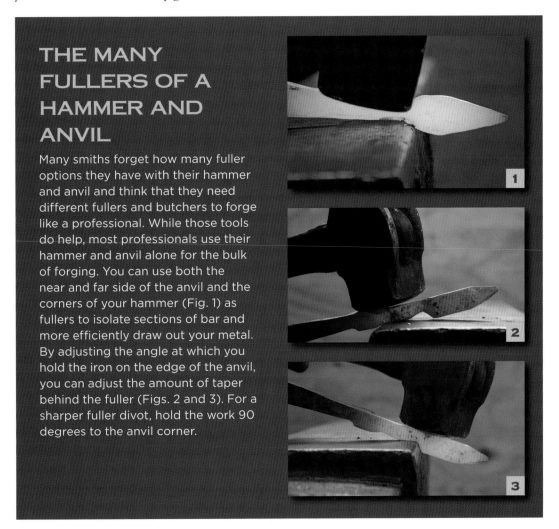

2.

SETTING
UP
SHOP

I see many new smiths stumble when setting up their shops, but now that you have a history of blacksmithing and, hopefully, a better understanding of how metal behaves, it should be easier to find tools for your shop. Don't worry if you can't find a perfect replica of the colonial blacksmith shop immortalized by Jefferson David Chalfant and other painters in the nineteenth and twentieth centuries; most blacksmiths through the ages had to make what they could with what they had, and that includes smiths in many parts of the world today.

Find a video online of smiths in developing countries if you want to see the blacksmithing spirit at its purest. These blacksmiths supply their towns with tools and household implements, using what they find; in fact, you may be surprised at the tools and items that they export to developed countries, too. They don't have time to look for the finest Peter Wright anvil, and neither do you—you want to get blacksmithing ASAP!

As I've mentioned, many new smiths become very frustrated, thinking that their work is inferior because of their tools. Remember that I started the book off with how metal moves and the importance of hammer control rather than jumping right to setting up shop to save you that heartache. If you skipped the first chapter, thinking you'd get to blacksmithing quicker, go back and read it now.

CHOOSING YOUR SITE AND ORGANIZING YOUR SMITHY

A properly set up smithy will ensure your safety and efficiency as a blacksmith. Everything should be organized and picked up off the ground to prevent you from tripping while holding a red-hot piece of iron or falling face-first into your forge fire. Because you will be using your forge, anvil, and vise most frequently, and because your hot metal will begin cooling as soon as it's out of the fire, these three tools should be situated close to each other. The best set-up is for the forge, anvil, and vise to form a triangle and to have a quench tub, also known as a slack tub, beside the forge, all within a few steps of each other to prevent losing too much heat while your iron is out of the fire. This is especially critical if you live in an area (like I do) in which temperatures are well below freezing for half of the year, and the environmental temperature quickly robs the heat from your steel. Your shop bench, where you will do your cold work, should be adequately lit to prevent eye fatigue while filing and fitting parts.

The actual location of a forge can be outside, inside a building, or anywhere in between. When deciding on where to set up, you need to consider a few things. First, and most important, is safety. Your shop, including floors and walls, needs to be fireproof and have proper ventilation whether you're building a shop or using an existing indoor location. You need fire extinguishers on hand to put out any type of fire because fires will happen. The building's roofing material should be fireproof, also, to prevent an ember from the chimney, if you have one for your forge, from starting a fire on your roof. On that note, if you use solid fuel, make sure your chimney is properly installed and up to code. As you can see, the number-one concern in a blacksmith shop is fire safety.

Second, think about your neighbors. If you are close to others, remember to do unto others as you would have done to you—be a good neighbor and look into soundproofing as well as how you will control disagreeable odors, such

A CLUTTER-FREE WORK AREA

No matter how cluttered the rest of your shop is, always keep your main work triangle between the forge, anvil, and vise clean and debris-free. To prevent fires, keep all flammable items out of the work area, particularly when forge welding, because the hot flux can spray in a radius of many feet.

To prevent heat loss in your metal and to be more efficient, your main work area needs to be within a few steps of the forge.

as smoke. If you can't build a soundproof shop, you may need to coordinate your smithing time with your neighbors' work schedules and your town's noise bylaws. How near you are to your neighbors also dictates your forge type because you can control the smoke by using charcoal or propane, rather than coal, as your fuel. Nothing shuts down your forge faster than an angry neighbor with the local police department on speed-dial.

Other factors to consider include the cost of building a shop, your local zoning bylaws, and the size of the items you plan to forge. If you have an outdoor work area, everything should be mobile so you can get it under shelter quickly during rainy or snowy weather, or you should have fireproof tarps handy. Tarps do double duty by providing shade when you are forging outside in the hot months.

Never underestimate the importance of flooring—as mentioned, it must be fireproof, and it should also be slightly soft to prevent fatigue when you stand on it all day. Traditionally, a forge floor was made of packed dirt, stone, or brick if the blacksmith didn't want his smithy to burn down too often. I have seen pictures in town history books of blacksmith shops with wooden floors, but such a picture is usually accompanied by a caption that details the numerous times that the shop had burned down.

While dirt is fireproof, nice to stand on, and cheap, it can be difficult to find and retrieve items that fall into it if you don't pack it down adequately. The dust from a dirt floor can also cause respiratory issues in some people. The other flooring option for today's shops is concrete, which is easier to clean and doesn't swallow up dropped items. On the other hand, concrete can cause lower back and leg fatigue. Many smiths with concrete floors use rubber anti-fatigue mats where they stand at the anvil, but you must purchase fire-resistant mats designed for welding; otherwise, a regular rubber mat can catch fire if a hot iron is dropped on it, creating an awful smoke that can be hard on your lungs. The best place to get suitable fire-resistant anti-fatigue mats is your local welding shop.

If your forge has cement flooring, a fire-resistant anti-fatigue mat will help prevent foot, leg, and back pain.

PROTECT YOUR SHOP

Whether building or renovating a shop, make sure that you have all of the proper permits and insurance. Nothing is worse than finding out after the fact that your insurance doesn't cover damage to your shop.

In my opinion, the absolute deadliest risk in a blacksmith shop is carbon monoxide poisoning. Carbon monoxide is an odorless gas that catches you unaware, which is why you must always have proper ventilation and a carbon monoxide detector in your forge. Ventilation will also save you if you accidentally throw galvanized steel into the forge because smoke from galvanized steel contains toxic amounts of zinc and has claimed blacksmiths' lives.

SHOP SAFETY

Blacksmithing is a dangerous activity, and it's likely that you will be hurt at some point in your forge, so your safety and the safety of everyone else in your forge need to be your main priorities. First and foremost, be sure to stay hydrated regardless of the temperature in the shop; blacksmithing is hot and sweaty work. In one blacksmithing class I took, another student almost fell into the fire when he collapsed from heat stroke, and we were in an air-conditioned building.

Speaking of water, also always ensure that you have enough water on hand to put out a fire quickly. This is extremely important if you are set up outside. On a bright day, grass fires can be hard to see until they get away from you, so always keep the ground around you dampened by sprinkling water from your quench tub. Keeping a dirt floor damp, but not muddy, will also help keep the dust down and pack the dirt into hardpan.

Every blacksmith shop should have a first-aid kit, including an eyewash station. Because iron can be as hot as to 1,000° Fahrenheit (538° Celsius) while appearing cold, you need to treat all metal in a blacksmith shop as hot enough to cause serious burns. Chips of metal can go flying from hammers and tools, bits of wire can dislodge from wire wheels, and scale from iron can end up in your eyes, requiring a way to wash out your eyes.

On the same topic, while forging, wear eye and hearing protection. A chip of metal in the eye can blind you, and the ringing of an anvil can cause permanent hearing loss over time. Also wear protective clothing made of flame-retardant material, such as wool, cotton, or linen. A leather blacksmith's apron is a good investment to protect both your body and your clothes from burns. Wear closed-toe boots, but never tuck your pants into bootlegs because hot scale from the iron can fall into your boots and burn you.

Regarding gloves, do not wear gloves other than on your chisel hand while hot punching or chiseling with nonhandled tools. Leather gloves are good insulators, but they shrink when hot; by the time you realize you're being burned, the gloves may be stuck to you, making the burn worse. On top of that, if the gloves aren't tight enough, they can get caught in any moving implements, such as a bench grinder, in your shop. If you treat every piece of iron as if it's hot, you will be much safer in the long run.

A glove on your hammer hand can also predispose you to repetitive-strain injuries because you will need to hold your hammer tighter, leading to increased strain on the tendons and ligaments because they—rather than your arm muscles—absorb the hammer recoil. When they are not tense, muscles are much more elastic than tendons and ligaments are, and they naturally stretch and shrink with normal use that would tear tendons and joint ligaments.

If the piece of metal is not long enough—generally less than 1 foot long—to safely handle without being burned, you will need to use tongs. New blacksmiths often don't properly fit the tongs to the metal, which is very dangerous. If the tongs do not fit the metal properly, the metal can fly off with

SAFETY TIPS

- Always use eye, ear, and, in some cases, face protection.
- Always wear flame-retardant fabrics such as cotton, wool, or leather.
- Always wear closed-toe boots with full pant legs pulled over the tops of your boots.
- Keep a first-aid kit and eye wash nearby.
- Have a fire extinguisher that is rated for all fire types, including oil and electrical.
- Keep your shop floor clean to prevent tripping.
- Never use loose-fitting tongs.
- Treat every piece of steel as hot enough to burn you.
- Don't use gloves while forging.
- Remember to stay hydrated.

tremendous force, leading to burns, cuts, bruises, broken bones, or eye damage. Properly fitted tongs hold the steel without any wiggling or slipping.

FINDING AND SETTING UP AN ANVIL

With what you've learned so far, you are ready to start looking for anvils, and maybe you have some ideas about places you might not have looked otherwise. For example, instead of spending hours looking for an antique anvil or spending much more money than your budget allows, as some other smiths may do, venture out to a scrap steel pile and begin forging sooner rather than later.

Before you begin looking for an anvil, you need to remember that the tools may make the craftsman, as the saying goes, but the blacksmith makes the tools. Throughout the ages, blacksmiths have used many different styles of anvils, but an anvil is simply something to hammer and bend hot iron on.

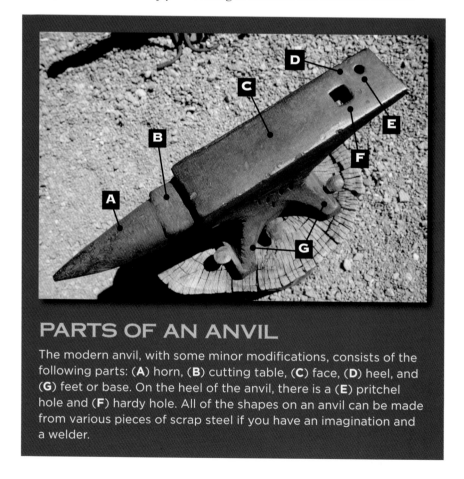

PARTS OF AN ANVIL

The modern anvil, with some minor modifications, consists of the following parts: (**A**) horn, (**B**) cutting table, (**C**) face, (**D**) heel, and (**G**) feet or base. On the heel of the anvil, there is a (**E**) pritchel hole and (**F**) hardy hole. All of the shapes on an anvil can be made from various pieces of scrap steel if you have an imagination and a welder.

Start with a simple block of steel, round one end of the face to create a horn, and cut a notch in the other end to let you drill hardy and pritchel holes—voilà, you have an anvil.

When choosing an anvil, look for three things: adequate working surface, mass, and rebound (see sidebar "What to Look For in an Anvil"). If a block of steel has those three qualities, its shape and its former uses do not matter. Look for an anvil that weighs more than 75 pounds for general forging and that is made of steel, not cast iron. Cast-iron anvils are known as anvil-shaped objects, or ASOs, by blacksmiths because they chip and break easily. Unfortunately, these are the easiest to find because they are sold by most welding and agriculture-supply stores. Spend your time and money at a scrap yard or farm auction rather than buying a cast-iron anvil.

GET TO KNOW THE HUNDREDWEIGHT SYSTEM

When you are shopping for antique anvils, you will notice that while some European models have the weight marked on the side in kilograms, most are measured in an old British weight system known as the hundredweight. There are three numbers, designating (1) the hundredweight; (2) the quarter hundredweight, which can only be 3 or less; and (3) the actual pounds, which is 27 or less. One hundredweight equals 112 pounds, and a quarter hundredweight equals 28 pounds.

For example, an anvil with a 1 1 1 on its side could be 111 kg if it is European or 141 pounds (112 + 28 + 1) in the hundredweight system.

Many beginning smiths think that a chunk of railway track makes a good substitute for a good London-pattern anvil. I find that railway track doesn't have enough mass under where you're hammering—unless it is stood on end—to work as a decent anvil under heavy forging with even a hand hammer. On top of that, railway track can be harder to find than a simple block of steel that would hold up better as an anvil.

Once you have an anvil, you will need a properly weighted and anchored anvil stand to keep it from bouncing. An anvil that weighs a few hundred pounds will work well sitting directly on the ground, but smaller anvils will bounce around too much for efficient forging. A stand can be anything from a tree stump to a fabricated metal stand as long as it adds weight to the anvil, is level and stable, and has a way to anchor the anvil to it. A quick search online will give you many different ideas.

BUILDING A FORGE

The forge is the heart of the blacksmith's shop, and it consists of three things: a fuel source, a place to hold the fire so you can stick your iron into it, and an air source to get the fire hot enough. It needs to produce adequate heat to make the steel malleable enough to work under hammer, which means 2,500° Fahrenheit (1,371° Celsius) to get the iron hot enough to forge-weld. If your forge doesn't reach welding temperatures, you are severely limiting the amount of work you can do and how efficiently you can do it; blacksmithing at near-welding temperatures is much easier than at cooler temperatures.

There are a few different fuel choices, and your fuel source determines which type of forge you need to build or buy. The most common fuels are solid fuels—coal and charcoal—and gas fuels, such as propane. A solid-fuel forge consists of a firepot with an air inlet that comes in from either the bottom or the side of the firepot. A propane forge is made of an insulated enclosure heated by a special torch. While you can make a propane forge cheaply, a solid-fuel forge is much easier to make, and solid fuel is often less expensive if you know where to look.

You can make an ash dump very simply from a threaded plumbing pipe found at your local hardware store.

Most blacksmiths in North America and Europe today feel that coal is the best fuel to forge with and would not touch charcoal. While coal creates a denser fire and generally works well, it does have some downsides. Proper blacksmithing coal is usually very difficult to find; on top of that, it produces more noxious smoke than charcoal or propane, so your neighbors might not be too appreciative of your new hobby.

DIY SIDE-DRAFT FORGE

You can use plumbing pipe to make a simple side-draft forge, but never use galvanized steel pipe because the galvanization can burn off and create toxic smoke.

WHAT IS A BOX BELLOW?

A box bellow is a simple, quiet, space-efficient way to supply air to your forge if you can't find a blower. All you need is some plywood, something to cut it with, and a drill to make holes. Build a set of joined boxes and create hinged valves and a handle as shown. To operate, simply push the handle back and forth to continuously supply air to your fire.

TO FORGE

What many smiths, particularly beginners, don't realize—because most books gloss over the fact—is that charcoal has been the primary fuel for blacksmithing throughout history. In fact, there are many professional smiths around the world today who still use charcoal. I have used charcoal as my primary fuel source because scrap wood, which would have been thrown away otherwise, was very plentiful on my farm. I find that a simple forge built for charcoal is best suited for beginner blacksmiths. If you can't find scrap wood to make into charcoal, you can purchase charcoal from any store that sells

You can make up your own firepot with scrap steel and a welder. For charcoal, the firepot needs to be around 5–5½ inches deep.

barbecue supplies. Just make sure that you buy actual charcoal, not the briquettes, because briquettes are designed for slow, low-temperature burning and do not get hot enough to use in a blacksmith forge.

As for forges, I have seen them made out of old large-truck brake drums, barbecue grills, wheelbarrows, oil drums, sheet metal, and cast iron; if they are made correctly, they work better than the riveter's forges that you find at most farm auctions. I started out with the latter but found that I was able to control my temperature and get my iron up to heat faster in a side-draft forge that I made out of an old 55-gallon oil drum and a piece of scrap pipe. To make your forge last longer, use ½-inch steel and insulate it with either firebrick or a fire-safe clay. You can find firebrick and different refractories at a fireplace store or online. You can also find many traditional fire-clay recipes that use bentonite clay, which is easily found in cheap unscented cat litter mixed with wood ash or sand, if you don't wish to purchase or can't find a commercial product.

If you are making a bottom-draft forge for charcoal, your firepot must be deeper than what you'd need for coal because charcoal is less dense than coal; this is why many smiths find that charcoal doesn't get hot enough. I feel that a firepot depth of 5–5½ inches works better for charcoal than the common coal depth of 3½–4 inches. With the increased depth, I have used softwood charcoal—which is the least dense of all types of charcoal and is considered unusable by many smiths—to forge-weld in a duck-nest style of tuyere. The walls of any solid-fuel forge should be at least ½-inch-thick metal to withstand the heat generated; any thinner, and it will burn through quickly.

MAKING YOUR OWN CHARCOAL

Making your own charcoal is less difficult than most people realize. There are many different ways to make it, some more efficient than others, but I found that simply starting a fire in a large drum and smothering it once it begins to char is very simple, if less efficient, than a charcoal retort. The key to making good charcoal is consistency of the thickness of your wood. I have found that between 1 and 2 inches works best as a thickness. As long as the pieces fit in the barrel, the length has no impact on how well it chars.

Using a barrel with holes around the bottom, start the wood on fire and let it burn until it becomes embers. Once the majority of the wood is embers, smother the fire by covering all the air holes and top, or douse it with water. Make sure the fire is completely out or it will continue to smoulder and you won't have any charcoal. Once it's cooled, sort out the partially uncharred pieces to be added to the barrel's next batch.

You can avoid making charcoal by using wood scraps directly in the forge. Make sure the pieces are about the same thickness as you would to make charcoal. Once the fire has burned to a nice charcoal fire, keep adding wood on to the outside of the fire and heating your iron in the charcoal. The only downside to this method is the excessive heat produced, which can be a good thing if it is cold in your shop.

A post vise is the best vise design for blacksmithing. It can be mounted to a weighted tire rim to be stable but mobile.

ESSENTIAL TOOLS
Vise

A post vise, also known as a blacksmith or leg vise, will make your life a lot easier, and you should be as particular about it as you are about your anvil. In fact, a proper post vise is one of the few things that a new blacksmith should either try to find as an antique or purchase new because they are difficult to make correctly. Depending on your area, there may be an abundance of post vises from farm auctions. If this is the case, grab one up. Most are in the 4–6-inch jaw-width range and can handle more work in a blacksmith shop than a larger machinist vise can; this is because the post supports the jaws of the vise, transferring the force of the hammer blows to the ground rather than causing the screw and mounting bolts to take all of the stress. A large machinist vise will suffice if you have no other options, but you should dedicate some time to looking for a proper post vise.

Hammers

Hammers are tools that have evolved differently in different regions of the world, but all types work well enough—if they are properly dressed—so that they can't be blamed for poor workmanship. Beginners can find good used hammers for very little money at garage sales, antique stores, or auctions.

HAMMERS

1. The face of your hammer should be slightly convex, with no sharp edges.
2. A flattened handle facilitates better accuracy and a more comfortable grip.
3. Top, left to right: A hardware-store cross-peen, a rounding hammer, and another style of cross-peen. Bottom: A ball-peen, or engineer's, hammer.

Sometimes a hammer will require a new handle, but handles are readily available at hardware stores. If you wish to purchase a new hammer, you can find decent hammers at most welding or farm-supply stores.

A blacksmith should have a selection of hammers of between 2 and 4 pounds, depending on how strong he or she is. To start out, you will want a cross-peen, a rounding hammer, and a ball-peen. Remember that hammer accuracy and your endurance make a bigger difference in how much forging you can do in a day than the weight of a hammer. However, initially overdoing the weight can cause muscle and tendon damage, so build up gradually to larger hammers.

Your hammer must be properly prepared before you use it; otherwise, you will end up with rough hammer marks and sore forearms from holding your hammer too tightly. With a grinder or sharpening stone, polish and round any sharp corners on the hammer's face and peen to prevent sharp hammer dings that make a forged piece look unfinished and amateur. A proper hammer face is slightly convex with smooth edges. As I previously mentioned, to help prevent carpal tunnel in your hammer wrist and tendonitis in your elbow, as well as to improve your accuracy, flatten the sides of the handle. A flattened handle lets you relax your grip and still keep the hammer from rolling in your hand.

CHISEL CUTTING EDGES

To get a proper 60-degree chisel edge on your cold chisel, use two large nuts and grind to match the bevel created between the two nuts.

FILE TIPS

Because hot iron cuts more easily, you can use dull files and save your sharper files for working on cold metal. Don't use your favorite file on hot steel because the heat will ruin the file's temper, making it soft and unable to cut cold metal properly.

Hand Tools

As you forge more items, your collection of specialized hand tools will grow. To start, you will want a collection of punches, chisels, files, and tongs; you need two sets of punches and chisels: one for use on hot metal and another for cold work. Punches and chisels used on hot metal can have finer points and angles than those used on cold metal because hot iron is easier to cut. You can forge your hand tools yourself, find them used at auctions or garage sales, or purchase them new. I don't recommend spending the money on purchasing new tools because you can often buy them used for much less money.

As you develop as a blacksmith, you will begin to collect tongs to hold hot steel of various sizes and shapes. You may luck out and find some tongs at good prices, but, like smiths of

SCROLLING PLIERS

An old pair of pliers, with the chrome removed, can make a quick pair of scrolling pliers to make curls. Simply disassemble, forge the jaws into round tapers, and rivet them back together.

HAND TOOLS

A. A good starter collection of tongs includes, from left to right: horseshoe tongs, flat bit tongs, bolt tongs, and another pair of flat bit tongs made from horse hoof nippers.
B. A collection of hand chisels, punches, and fullers, including a decorative eye and nose punch for the horse head in the projects.
C. These are examples of handled tools. From left to right: butcher, hot cut, and slot punch.

old, you can make any tongs that you need. I will cover making tongs later in the book.

Other hand tools are used to isolate and smooth metal. *Butchers* and *fullers* are special types of dull chisels used to isolate sections of metal, while *flatteners* and *set hammers* are used to smooth the metal. To set shoulders on tenons, you will need to make *monkey tools*, which are simply rods of various diameters with different-sized holes drilled in them. The tenon fits snugly in the drilled hole, and you hammer the monkey tool down onto the shoulder to square it up.

An old ball-peen hammer can be forged into an easy handled chisel, fuller, or punch. Just forge the ball-peen out and heat-treat it.

Measuring Devices

You will also want a way to measure curves and bends from your layout. A long piece of soft wire or string will be a lifesaver when it comes to measuring final dimensions of scrolls and other curves. Some smiths make a measuring wheel, which is simply a pizza cutter with graduations marked into it so that you can count the number of

A nail header hanging from the first nail it created.

revolutions it takes to follow the piece in question, but it is not necessary. I personally use a cheap sewing tape on cold metal and wire on hot metal. Other handy measuring devices include calipers for measuring the diameter of stock and metal squares to ensure that you are perfectly square when it matters.

Anvil Tools

Working alone or with a striker, you will rely on many different anvil tools. The anvil tools you will use most frequently are a hardy, various fullers, swages, and a chiseling plate. A hardy is simply a chisel with a square shank that sits in the hardy hole of the anvil to cut hot metal. There are two styles of hardy that I have come across in my years of blacksmithing: most smiths use a straight-edged hardy, but a growing number are using a curved-edge hardy. I find that while the straight-edged hardy works well, a curved edge bites into a bar better, preventing it from skipping and making extra cut marks that you'll need to hide. On the other hand, a straight-edged hardy is less likely to end up getting damaged by your hammer because the center of the hardy has cut through the iron before its edges. For the beginner smith, either type of cutting edge works, but make sure it has a shoulder to keep it from getting wedged into the hardy hole if you cut cooler metal with it.

Some smiths will make their hardies with tapered shafts and no shoulders so that they are more solid when in use, but I don't recommend this style for new smiths because many anvils have the heel broken off from a hardy wedging in with use, and new smiths are more likely to try to cut cold metal with their hardies. If you cut cold steel with a shoulderless hardy, you risk breaking the heel of your anvil off. All other hardy tools need to have a shoulder—the cutoff is the only hardy tool that can get away without a shoulder because it is cutting through hot metal. Any others could break the heel off your anvil if the shank is too large for the anvil's hardy hole.

Use a hardy to cut off hot metal.

ANVIL TOOLS

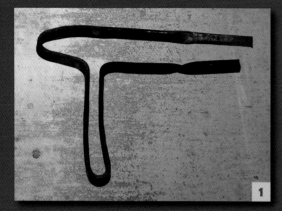

1. You will learn how to make a spring fuller later in this book.
2. Another style of fuller that can be made with access to a welder.
3. Investing in a welder will make your life much easier when it comes to making anvil tools. From left to right: A saddle anvil for forging branches and forks, a cupping or dishing swage, bending forks, and a V swage made from angle iron.

The easiest way to cut off metal when it is hot is to use your hardy in the anvil, but you must take care to not damage your hammer face when using your hardy. To properly use the hardy, begin by hammering straight down onto the cutting edge. If you are cutting square metal, continue by rotating to the corner, hammering, rotating to the flat, hammering, and repeating until you are almost through the metal. At this point, you can either break it off with a pair of tongs or continue to cut it off with hammer blows off to the side of the cutting edge. To cut round steel, simply continue to rotate the rod while overlapping the cuts.

A single smith should also have a collection of spring fullers to fuller and isolate metal similar to the matched anvil and top tools that would be used with a striker. The simple spring fuller is a project presented later in the book. It can be made with differing widths of bar and round rod to make different sized fullers. It is designed to be wedged solidly into the hardy with a wooden wedge.

3

STEEL TYPES
AND
HEAT-TREATING

As I've alluded to, modern steel is not composed of pure iron molecules; rather, it is an alloy containing varying levels of carbon and other trace elements, such as chromium, molybdenum, and vanadium, to name a few. Beginning blacksmiths should stick to low-carbon mild steel because it is the most forgiving of all the steel types when it comes to the forging temperature. It is also the most malleable alloy and the easiest to forge. Of course, as you progress, you will want to expand into other steels for tools and blades, so you should learn how to properly heat-treat metal, particularly if you are using scrap iron. Scrap iron can be anything from low-carbon mild steel to tool-steel alloys that require special heat-treating.

To save money, you can often salvage metal for free or almost free if you know where to look. If you are planning on paying for scrap, always compare prices between scrap steel and known alloys from a steel supplier. Often, there

TREAT IT RIGHT

If you purchase new steel, always ask for the heat-treating information for that alloy so that you can properly heat-treat the steel after forging.

is not very much of a price difference and, particularly if you are serious about making and selling good-quality tools and knives, it is better to know what you are working with rather than having to guess the alloy—and then have your tools break because you guessed wrong. In this scenario, the best case is that you spent a lot of time forging a tool that broke before you sold it; the worst case is that the tool injured or even killed someone. That being said, it is very rewarding to be able to make something with your own hands from iron you recycled for free; if properly forged and heat-treated, such metal can make very serviceable tools for you to use and sell.

SALVAGING STEEL

As I've said, salvaging steel can be a very rewarding endeavor for a blacksmith. You find a piece of metal, perform some tests to determine what it is best suited for, and make it into something completely different. On the other hand, it can be frustrating if the piece breaks because of a fault in the steel. More often, though, breakage is due to the smith's heat control or because the smith guessed the alloy wrong, not because of the steel itself. The best tip for forging salvaged steel is to treat all of it as high-carbon steel—never forge it too cold and, most importantly, never fully quench it in water unless you need to harden it and you have already tested it.

COMMON STEEL'S GENERALIZED CARBON LEVELS

Never assume that a steel is what you think it might be based on the various junkyard steel charts available. Companies often change their steel alloys over time—what used to be low-carbon may now be high-carbon, and vice versa. Always test the metal's hardenability before you decide to use it for a project. Because of this variability, I will only go into generalizations of the types of steel—it is best to think like a metallurgist and consider which type of steel best suits the function.

Low carbon—Non-weight-bearing, non-tool steels; examples are rods or angle iron from old railings, barbecue grill stands, and the like.

Medium carbon—Many agriculture and automotive parts are made of medium-carbon steel for durability; examples are axles, coil springs, struts, tie rods, harrows, and plow discs.

High carbon—Many cutting instruments, such as saw blades, jackhammer and drill bits, hay-rake tines, punches, chisels, and leaf springs are made of high-carbon steel.

TIPS FOR BETTER SALVAGING

Salvaging steel is an art, and it can lead to frustration if the steel breaks after you've spent hours forging it into something useful. If everything goes right, though, salvaging can be one of the most satisfying experiences for a new blacksmith, and you'll become hooked. Recycling steel and repairing old iron ties you to all of the blacksmiths throughout history. Keep the following tips in the back of your mind while scrounging to minimize the chances of having a critical failure because of the salvaged steel.

- Work your metal hot to minimize cracking and hardening the metal.

- Match the metal's new use to its previous use. If it was used in a high-strength or torsion part, use it as tool steel, not mild steel (this is where most scrap iron fails because the smith quenches it as if it were mild steel, and it becomes brittle).

- Always have known samples to compare with when spark testing; every grinder will throw slightly different sparks for a given steel.

- Wire-brush the steel and look for cracks and weak points.

- Always allow your piece to cool slowly. Never quench unknown steel in water unless you have tested its hardenability.

- Start with a slower quench and softer temper when you first use a piece of salvaged steel; this will prevent chipping and let you reforge the tool if needed. Once it chips or shatters, you will need to start over, and you may run out of similar stock before you get the heat-treating right.

When salvaging steel to recycle in your forge, always try to match uses. While this won't prevent issues during forging, it will help give you an idea of the minimum forging temperature, how the metal will handle various quenches, and what type of project it should be used for. For example, you can expect that a vehicle axle would be good for making into hammers because axles are meant to take abuse and maintain their strength and shape, and most axles are made from 4140 alloy steel. At the same time, if you misjudge what type of steel it is, you could end up with a broken coat hook if you forge it down into a dark orange color or quench it in water.

Regardless of whether you are using purchased steel of a known alloy or scrap steel, you need to know how steel works to best select the right type of steel for a given project as well as to understand how to properly heat and cool it. As you progress into toolmaking, it is in your best interest to read more on metallurgy.

HEAT-TREATING

Before you start forging scrap metal, you need to understand how steel changes with heat and cooling. Steel has the miraculous property of hardening when cooled quickly, but at the expense of becoming brittle. How hard it can get changes primarily with the amount of carbon in the steel. For most blacksmiths, the carbon is the critical molecule in steel. As the amount of carbon increases, the steel's hardening ability increases as well. Mild steel has very little carbon and can only be hardened in certain instances, which makes it great to use in most household items but not useful for tools or knives. If steel with greater than 0.4% or 40 points of carbon is hardened without tempering, it can become brittle and is often the reason why some smiths curse using salvaged steel: it breaks because they didn't realize they were forging higher carbon steel and treated it like mild steel.

Heat-treating steel requires three steps: annealing/normalizing, hardening, and tempering. The following

SPARK-TESTING UNKNOWN STEEL

A common test to determine the carbon content of unknown steel is to see what the sparks look like when the steel is ground. Basically, the more the sparks turn into a starburst, the higher the carbon content. Make sure to test the metal against known higher carbon and mild-steel pieces on the same grinder because every grinder can throw sparks differently.

ADDING CARBON TO STEEL

You can add carbon to the surface of mild steel to make it slightly hardenable. There are commercial products available, but blacksmiths traditionally added carbon by surrounding the iron in ground-up bone or hooves in a metal box and keeping it heated to cherry red for a day or two. This process let the carbon migrate into the steel so that the skin could be hardened.

CARBON CONTENT

Carbon content in steel is commonly measured in either the percent system or point system. Because cast iron has the highest carbon content, at 1 percent, other steels have between 0 and 1 percent. The point system is slightly different in that cast iron has 100 points of carbon and other steels have between 0 and 100 points.

is the simplified, down-and-dirty version of heat-treating that will serve you well most of the time. I encourage any blacksmith to learn metallurgy because it will help you make tools and other projects that will last longer, work better, and save you the frustration of watching the knife or chisel you just made shatter.

Annealing/ Normalizing Steel

To normalize or anneal steel, slowly cool it after forging to let the steel's grain structure relax and refine. This reduces the chances that the steel will shatter when it is hardened. For the basic blacksmith, normalizing means to slowly cool the steel near the fire, while annealing means to very slowly cool the steel

THICK, THEN THIN

Always leave your pieces slightly oversized by about $\frac{1}{32}$–$\frac{1}{16}$ inch to allow for carbon loss from the outer surface during heat-treating. This carbon loss will reduce how hard the very edge can get. Thinner steel is also more prone to warping and breaking during hardening. Remember the old blacksmith quote: "If you will a blade to win, first forge thick, and then grind thin."

in an insulating material such as ash or vermiculite from a garden-supply store. Because this is the simplified version of metallurgy for beginning blacksmiths, I won't go into detail about how the speed of cooling changes the grain structure, but suffice it to say that normalizing neither softens nor refines the grain of the steel as much as full annealing does. Normalizing is fine for lower carbon steel, but you should fully anneal any steel with greater than 0.6% or 60 points of carbon to be safe. If in doubt, fully annealing a piece of steel will do no harm other than taking a little more time.

To normalize or anneal steel, first heat it to slightly above that alloy's critical temperature. At that point, the steel grain is as hard as it can be, but if you go too hot, the grain structure can become too large, making it more prone to fracturing. Every alloy has a different critical temperature that you should try to figure out, but for most items that the beginner smith produces, simply wait until a magnet won't stick to the steel. If you wish to move into bladesmithing, you will need to be more precise about temperatures.

When you reach the desired temperature, remove the metal from the fire and either place it along the edge of the fire or bury it in ash or vermiculite, depending on whether you want to normalize or anneal the steel. After it is cooled, you will be able to easily file the steel into shape without prematurely dulling your file. If you can't file it, you need to reanneal it.

Hardening

Once your tool is softened, you need to harden it. By cooling the iron quickly, the carbon molecules in the steel cannot precipitate out as they would normally, and you lock the steel grain into its hardest form: martensite.

HARDENABILITY TEST

Test whether a steel is hardenable by hardening it in different quenching liquids and testing it with a sharp file; if the file skates across the steel, it is hardenable steel.

How you quench the steel is very important. Many blacksmiths quench their steel in three different liquids—salt water, water, and oil—although some don't bother with salt water. These different quenchants cool the steel at different speeds due to their thermal conductivity. Salt water cools the fastest and can make higher-carbon steel very brittle due to thermal shock; water and oil cool progressively more slowly. Steel is just like Goldilocks—the rate of cooling needs to be just right. Depending on the steel you use, it may need to be cooled in either water or oil to ensure that it becomes hard but not too brittle. If you begin making knives in which warping is an issue, heat the quenchant to slow its rate of cooling and minimize how much it shocks the steel.

Tempering

After you have quenched your steel, you need to relax the martensite to make the steel less brittle by tempering it. To do this, heat the steel slowly and quench when it reaches a certain temperature. The temperature you bring it

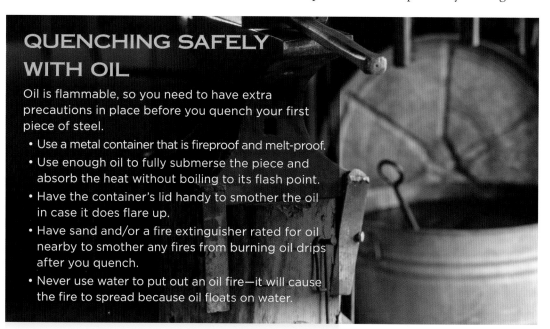

QUENCHING SAFELY WITH OIL

Oil is flammable, so you need to have extra precautions in place before you quench your first piece of steel.

- Use a metal container that is fireproof and melt-proof.
- Use enough oil to fully submerse the piece and absorb the heat without boiling to its flash point.
- Have the container's lid handy to smother the oil in case it does flare up.
- Have sand and/or a fire extinguisher rated for oil nearby to smother any fires from burning oil drips after you quench.
- Never use water to put out an oil fire—it will cause the fire to spread because oil floats on water.

TEMPERING COLORS

Use oxide colors to determine the steel's temperature range. Starting at yellow and progressing to blue, the temperatures will run as shown as the steel heats up and softens.

Yellow (350–400°F): Hardest temper, used for cutting tools such as wood-carving chisels and slicing knives. Anything that needs a hard, strong edge at the expense of a higher chance of chipping or cracking. The steel will not flex much without breaking.

Brown (400–450°F): Harder intermediate temper for tools that need more flex, such as chopping blades, or that need more shock absorption, such as hammer faces, struck cold chisels, and punches.

Purple (450–500°F): Softer intermediate temper, allowing for even more flex without breaking. I like to use this temper for tools that will be used for prying, such as crowbars and claw-hammer claws. If the item will be used for a lot of hard prying, I will go to a full blue temper.

Blue (500–550°F): This is known as a "spring temper," and it allows the maximum amount of flexibility that needs some toughness. It is most often used on, not surprisingly, springs. If you go any cooler, you can create stress fractures in the steel, often showing up as multiple small cracks.

to will determine the toughness of the steel—as the steel becomes hotter, it becomes softer and therefore less brittle.

You can temper the metal by heating the whole piece either in an oven or over the fire, or you can temper it progressively by heating it from a part that needs to be softer. "Letting the colors run," as progressive tempering is called, is best for tools such as punches and chisels in which you want one end to be hard and the other completely soft. To do this, you simply heat the tool up to critical temperature and then quench only the tip of the tool to harden it. After hardening, quickly polish the tip of the tool and watch for the proper temper color to show up. As soon as the temper color you want reaches the tip of the tool, quench the whole tool. This leaves the working end hard and properly tempered while gradually softening to the end you hit with the hammer and so prevents breaking and chipping at a sudden change of hardness.

BASIC
BLACKSMITHING
TECHNIQUES

I t is said that blacksmithing is easy to learn but impossible to master. Compared to other arts and trades, there are very few techniques to learn, but you will always be improving them as long as you forge iron. The basic techniques in this chapter will not seem very basic when you first start out, but you will use them in every project your make.

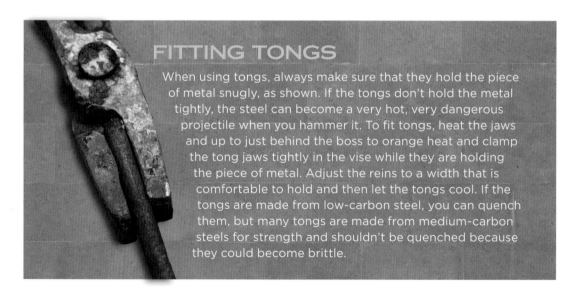

FITTING TONGS

When using tongs, always make sure that they hold the piece of metal snugly, as shown. If the tongs don't hold the metal tightly, the steel can become a very hot, very dangerous projectile when you hammer it. To fit tongs, heat the jaws and up to just behind the boss to orange heat and clamp the tong jaws tightly in the vise while they are holding the piece of metal. Adjust the reins to a width that is comfortable to hold and then let the tongs cool. If the tongs are made from low-carbon steel, you can quench them, but many tongs are made from medium-carbon steels for strength and shouldn't be quenched because they could become brittle.

HAMMER BLOWS

Before you begin practicing blacksmithing, you need to know the different types of hammer blows, which vary based on how you are hammering the metal in relation to the anvil: full-faced, half-faced, and shearing. All of these blows can be done on any part or edge of the anvil as well as at any angle to isolate and fuller the steel.

With a full-faced blow, the metal is fully pinched between the hammer and anvil; this blow is used to taper, draw, and smooth the steel. With a half-faced blow, the metal is forged only partially on the anvil to either create a shoulder in the metal or to protect another area from being forged. A half-faced blow can also be used to fuller and isolate metal more efficiently. I encourage you to practice using the edge of your hammer and anvil for half-faced blows to maximize your efficiency. With a shearing blow, your hammer does not hit the anvil at all. It is used to bend steel and can be done over the edge of the anvil, the horn, or any mandrel or dishing tool.

FIRE CONTROL

The main difference between metal fabrication and blacksmithing is the forge, which heats the metal to temperatures that make it moldable—with a little persuasion from the hammer and anvil. To heat the metal properly, you need to be able to control your fire—too little heat, and the metal will be too hard to forge and more prone to cracking; too much heat, and you will burn or melt the metal, rendering it useless. The amount of oxygen in your fire is also important because too much air can lead to increased oxide, also known as *scale*, formation.

Your fire temperature and oxygen control play a huge role in how easily you forge-weld—even more so than the actual fuel you use or how clean your fire is—by minimizing the amount of oxide formation. I have forge-welded in softwood charcoal fires and coal fires

Full-faced blow.

Full-faced blow.

Half-faced blow.

Shearing blow.

LIFT THE FIRE

Your fire will settle often and block the airflow, particularly if you are using coal. Refresh the fire by using your poker to lift the fire off the tuyere and break it up to increase the amount of air entering the fire; this will help keep it burning brightly.

with actual dirt and shale in the coal; it can be done if you pay attention to the amount of oxygen the iron is exposed to while you heat it up.

How easily your forge fire starts depends on the fuel source. Charcoal and propane are very easy to light, requiring only a piece of paper. Coal is much more difficult to start, especially if you haven't got any coke. Coke is coal that has had the impurities burned out of it; coke is to coal like charcoal is to wood. To light a coal fire, you will need more starter fuel than you do for charcoal, but the steps are the same for both fuels. Bundle up five to ten pieces of newspaper into a tight wad or use some kindling or charcoal to create a small fire in the firepot.

Personally, I find that simply starting a small wood fire works best. Slowly add air to get the starter fire burning well and then begin piling the coke from a previous session around the fire. If you don't have coke, use more paper or wood and pile the coal around the fire. Slowly increase the amount of air that you are feeding the fire until it is burning bright and flames are visible. Coal will smoke a lot, so you'll need a good chimney to clear the air and prevent lung damage. Charcoal will smoke, too, but to a much lesser extent. The main issue with charcoal is the embers, so always have water available to put out stray fires and a cap on your chimney to reduce the number of embers that are released into the air. Remember that you need fire-resistant roofing material if you will be using any type of solid fuel.

Once your fire is started, you need to contain it with strategic use of water. This will keep you from using more fuel than needed. If you are using coal, dampening the dust will let it coke up properly rather than sifting down and blocking the tuyere, thus killing your fire. Always keep the area outside the metal you are heating damp to prevent the fire from spreading, which wastes fuel. Slowly feed your fire from the outside edges to let your fuel burn properly, particularly when you are using coal. Green coal will stick to your metal, becoming a major nuisance and often finding the wrong place to fall off, causing burns. Coal also doesn't burn to create heat until it becomes coke. By feeding the fire from the outside, you burn out the impurities to let the coal coke up properly and stay hot.

The fire itself is made up of three layers, starting at the tuyere: the reducing, neutral, and oxidizing layers. The deeper into the fire you get, closer to the air source, the more oxygen is present. In the reducing, or *carburizing*, layer, there is more fuel than oxygen. If the iron soaks long enough in this layer, you can create blister steel as more carbon is added to the steel. The neutral layer is balanced between fuel and oxygen. The neutral layer is the best layer to heat your iron in. It not only has the least impact on your steel's composition, but it also is the hottest layer and the easiest to forge-weld in. In the oxidizing layer, there is excessive oxygen, and you will experience much more scaling on your steel. If your steel soaks too long in the oxidizing layer, you can also experience some loss of carbon in your steel. For most of your projects, the amount of carbon lost or gained will have no impact on the end result; however, if you begin forging more specific tools that have a difficult life and are expected to hold up under intense strain, such as knives, tow hooks, or jackhammer bits, it can make a difference.

FUEL TYPES: PROS AND CONS

Charcoal—*Pros:* Easy to make or find; easy to light and use; not as much smoke. *Cons:* Embers can burn and start fires; you use more fuel; it can become expensive over time if you have to purchase barbecue lump charcoal; fire spreads quickly, leading to more fuel loss if you don't keep edges damp; needs to be put out when not in use to prevent it from continuing to smolder until it is all used up. *Note:* Make sure that you are using actual charcoal, not barbecue briquettes.

Coal—*Pros:* Very hot per volume of fuel, you need less fuel; fire doesn't spread quickly; it will stop burning when the air supply is cut off; doesn't produce embers. *Cons:* Produces large amounts of noxious smoke while still green; good-quality blacksmithing coal can be difficult to find and expensive; fire is harder to start. *Note:* Make sure that the coal burns hot enough.

Propane—*Pros:* No smoke; no risk of embers or sparks; easily found; burns at a constant temperature; doesn't require constant monitoring once set. *Cons:* Forge is more difficult to make, it can become expensive to buy fuel, there is a risk of explosion; death from carbon monoxide is still a possibility even without smoke.

TEMPERATURE BY COLORS

The great thing about steel is that you can tell its approximate temperature by its color. Forging temperatures range between 1,600 and 2,300° Fahrenheit. The bright yellow in the center of the heated portion is at a high welding temperature (2,300°F), while the tip is at a lower forging temperature (1,600°F). Bright orange is the coolest at which you should ever forge steel; any cooler, and you can create stress fractures in the steel, often showing up as multiple small cracks.

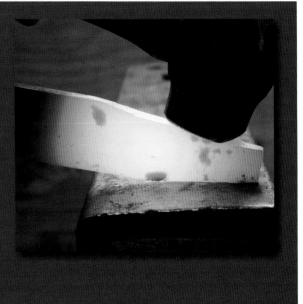

FIRE-MANAGEMENT TOOLS

To properly manage your fire, you need a few tools. A coal scoop, water ladle, shovel, poker, and rake will keep your fire in tip-top shape. It helps to have a differently designed handle on each so that you can easily tell your tools apart.

FORGING

Now that you are able to get your metal hot enough, it is time to begin forging.

Drawing and Tapering

Drawing and tapering to lengthen and thin steel are the most common techniques you will use as a blacksmith. When you draw steel out, you use flat hammer blows to make a section of steel uniformly thinner and longer. Tapering uses angled hammer blows, often with the piece of metal held at an angle to the face of the anvil, to narrow a section of metal on an angle.

To most effectively draw steel out, you can use a fuller to move the metal in the direction you want, rather than spreading. Holding the steel flat on the anvil, switch between fullering the steel and flattening the ridges you create until the steel is the thickness you want. Because the metal will inevitably spread while lengthening, dress the width of the bar to your desired width by holding it on edge, flat against the anvil, and hammering parallel to the anvil face. Correct the width of the bar as you go, before it gets too thin, to prevent folding the metal over. You can use a cross-peen, hammer corner, anvil corner, or anvil horn as a fuller to do this.

Tapering lets you narrow the steel, giving a sense of dynamics and fluidity to your work. Mild tapering is also a way to quickly draw down steel by creating a fuller effect with the hammer and anvil corners. To create a two-sided taper, as you would for a chisel, hold it at an angle on the edge of the anvil and match the angle with the hammer face. You can use either the far side or near side of the anvil to taper, depending on where in the steel you want the anvil.

You can make a one-sided taper by tilting the hammer but keeping the metal flat against the anvil face. If you want to create a point on a rod, also known as a four-sided taper, simply flip the steel 90 degrees between hits.

Use flat hammer blows to draw steel out.

To create a two-sided taper, hold the metal at an angle to the anvil and match the angle with your hammer.

Create a single taper by holding the iron flat on the anvil and angling your hammer blow.

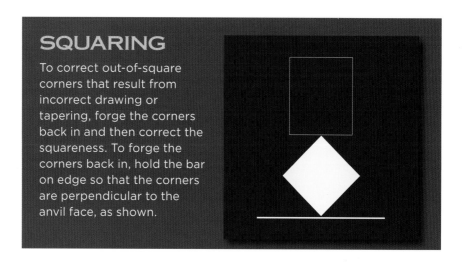

SQUARING

To correct out-of-square corners that result from incorrect drawing or tapering, forge the corners back in and then correct the squareness. To forge the corners back in, hold the bar on edge so that the corners are perpendicular to the anvil face, as shown.

Rounding

Many new blacksmiths struggle with rounding steel because they don't follow the proper steps. When they try to taper a rod round without squaring it first, they end up flattening the tip into an oval and then completely flat, twisting the metal and leading to cold shuts and cracks as it twists tighter and tighter. Eventually, the tip will split in a spiral once it is thin enough. To properly round steel, you first need to square the rod by turning it 90 degrees between hammer strikes. From there, you need to make the rod into an octagon by knocking the corners back into themselves. Make sure that the corners are in line, perpendicular to the anvil face; otherwise, the rod will twist, weakening the steel by creating a cold shut. Continue knocking the corners back until there are no more corners to hit.

Always go from square to octagon and then round, even when it's a round rod.

Fullering to Isolate and Spread Steel

Just like you can pinch off modeling clay, you can isolate and spread iron as well. To do that, you will need to fuller the metal. A fuller is simply a wedge or rounded tool with different radiuses that forces the metal perpendicular to the wedge.

Your hammer and anvil are the fullers you will use the most; they are the easiest to find and require no setup time. A good blacksmith knows all of the shapes he can get out of his hammer and anvil before he reaches for a fuller. That being said, spring fullers and guillotine tools can make your life a lot easier if you need to make many duplicates. Just don't jump to making jigs and tools before practicing your hammer control because you won't reach your full potential as a blacksmith as quickly as you would by using your hammer and anvil as your fullers.

Upsetting

There is a blacksmith joke that "upsetting" is called "upsetting" because it is. When you start upsetting metal back into itself to make it larger, it can be very frustrating because the metal wants to fold over constantly.

When upsetting, a good rule of thumb is to heat only 1½–2 times the width or diameter of the material to minimize buckling. If the metal does start to fold, you need to straighten it, or else the fold will become a cold shut and weaken the piece at that crack.

There are a few different methods to upset steel depending on the desired end result and the size of the metal. For smaller rods, I find it easier to begin the upset in the vise because it is hard to get the required short heat to prevent buckling. With a larger piece, you can either hammer it down on your anvil or on an upsetting block with a hammer. If it is heavy enough, you can simply pound it down, using its own mass as a hammer. In some instances, such as making a hardy tool, you can upset a section to create a shoulder that catches the end you want to further upset, letting you enlarge just part of the tool, which saves time because you don't have to draw the shank down from a larger rod.

FULLERING

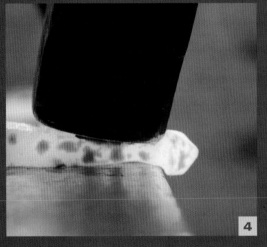

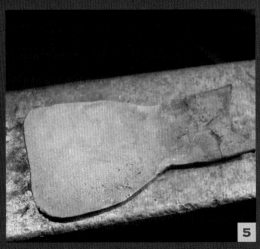

1. You can use the corner of your anvil as a fuller to isolate metal.

2. The cross-peen hammer is another fuller that you can use to be a more efficient blacksmith.

3. A simple homemade fuller guillotine will isolate both sides of the metal quickly and consistently.

4. Both the cross-peen and the corners of your hammer face can be used as fullers.

5. With a fuller, you can easily spread metal wider.

UPSETTING

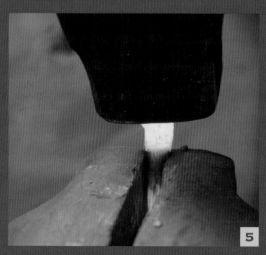

1. One way to upset is to hammer the bar into itself over the anvil.

2. You can hammer the hot end into a hardy swage to help round the end.

3. A nail head is the perfect example of an upset in use.

4. Your hammer blow will dissipate as it travels into the metal, upsetting the metal more at the point of contact than below.

5. To help keep smaller diameter metal from bending while being upset, you can clamp it in the vise with approximately 1½–2 times the diameter above the jaws.

MARKING YOUR METAL

It is very difficult to see chalk on hot metal, so you need to mark your distances with a punch or slight chisel notch. Always mark the inside of a bend to hide the mark.

It is important to note that the force of the blow dissipates as it gets further into the bar because of the physics of inertia and friction between the iron molecules. You need to take this into account when planning how you are going to upset a bar. You can change how a piece is upset by three factors: heat, force/weight of the blow, and distance from the blow.

First, how you heat the bar impacts where most of the upsetting will occur because hotter metal is more malleable and moves more easily. The cooled portion of the steel opposite your blow becomes an anvil, forcing the hot metal outward.

Next, the force of your blow will impact how much energy there is to be transferred down the steel. Any time you upset a section of metal, it will be more upset closer to the blow and slowly taper off as it gets farther away. Depending on the look you want to achieve, this can be used to your advantage. To overcome the riveting of the end and make the upset deeper, you need to increase the force of your blow, generally by using a larger hammer and hitting harder.

Because the energy is lost the farther into the bar you go, the distance between the blow and the heated portion affects how easily you upset the section you wish to be upset. For example, you would be wasting your time if you hammered on the cool end of a long bar to upset the other end because the energy would be absorbed by the bar. To efficiently upset a piece of metal, regardless of whether it is at an end or within the bar, always heat the section you want to upset to the hottest temperature and have the force of the blow start as close to the upset as possible.

Upsetting also lets you punch larger holes in bars than you could normally without sacrificing the strength of the bar. By combining slot punching or slitting a bar and upsetting the hole to open up the hole before drifting it to size, you lose much less sidewall than you would by simply punching or drilling it at the exact size of hole you want.

Bending and Scrolling

Bending and scrolling metal are both the easiest and most difficult techniques you will encounter as a blacksmith. The technique itself is easy to learn, but to make it look good consistently, especially if you need bends or scrolls to

BENDING

1

2

3

1. Begin the bend over the far side of the anvil with a shearing blow.

2. Finish the 90-degree bend by hammering the edge over the side of the anvil as shown.

3. Bend over the horn to create wider radius bends and rings.

match each other, is the toughest thing to master. It takes a critical eye and practiced hand to control the bend the way professional blacksmiths can.

To bend steel sharply, use shearing blows off the edge of the anvil. You can use the near side of the anvil in some cases, such as when you need an exact length bent. By lining up to a chalk mark on the anvil face and then tilting your tong hand down to lift the end off the anvil face at the near side of the anvil, you can hammer the tip down to start the bend without needing to mar the piece with a center punch mark.

When bending, you will find that the metal will naturally flex and bend because of the inertia to get the end of the bar moving. To straighten the legs of the bend, lightly hammer directly against the side and face of the anvil, taking care to not thin the metal as you hammer it. Normal 90-degree bends are actually a tight radius, not a perfectly square corner. Getting a perfectly square corner is much more time consuming. It is vexing to blacksmiths that customers are willing to pay extra for a simple, quick twist yet expect a perfectly square corner that takes twice as long to be cheap.

SMOOTH SCROLLS

While most smiths will tell you that you need to practice your hammer control to improve your scrolls, they often neglect the topic of heat control. The easiest way to get a smooth curve to your metal is to ensure that the whole section that you are scrolling is consistently heated and that the heat tapers off smoothly at the end of the scroll. By tapering the heat off the ends, you create a naturally smooth resistance to bending, which helps smooth out your scrolls. Never try to scroll anything where the scroll passes into a cold portion of the bar because you will end up with a sharp bend where the heat stops.

TYPES OF SCROLLS

There are a many different types of scroll finials, but the following are the most common. Experiment with adding leaves or making other styles of side scrolls.

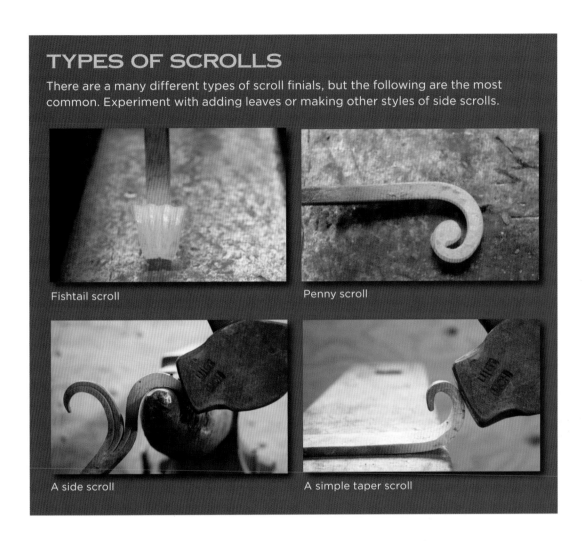

Fishtail scroll

Penny scroll

A side scroll

A simple taper scroll

For a perfectly square corner, you need to upset metal into the corner to forge it square. Because of this, leave a little extra in your measurements to allow for the upsetting. For most light work, ¼ of an inch generally works, but you should figure out the math to be more exact for the dimensions of the metal that you are using. Remember that metal is a set volume: you can determine the volume you need for the end piece and calculate the starting material that you need. When in doubt, leave both ends a little longer and cut off the excess when finished.

Begin a square corner by bending the metal 90 degrees, as you would to make a standard bend. To upset the metal into the corner, hold the piece so that the leg you are not holding is facing up, perpendicular to the anvil face, and hammer directly down on the leg. Straighten the leg as it buckles. To upset

SQUARE CORNER

1. Upset metal back into the corner to let you forge it square.

2. Continue to upset the other leg of the corner over the edge of the anvil. Notice the gap between the bar and side of the anvil. If you don't have a gap, you'll end up drawing the rod out rather than upsetting the corner.

3. Once you have upset both legs of the corner, you can forge the corner square.

the other side of the bend, hold it over the far edge of the anvil and hammer the corner back toward your tong hand. Make sure there is a gap between the corner of the stock and the anvil to allow for upsetting. Once you have upset enough metal at the corner, square it up by hammering it on the far edge of the anvil with half-faced blows on the anvil face and side. Square up the sides of the bend because they will swell with the upsetting.

To get a square corner inside and out, you need a square edge on the anvil or a block of steel with a square edge to hammer over when squaring up the outside of the corner; otherwise, if you use only your rounded anvil corner, you will have a sharp corner on the outside, but the inside radius will match the anvil edge.

Scrolls are the quintessential iron decoration, showing the fluidity that you can achieve with iron while maintaining its strength. Scrolls lend a beautiful juxtaposition to projects that only iron can achieve. To achieve that beauty, a smith needs to produce a smooth curve without any sharp transitions. The secret to achieving this is a consistent heat throughout the bend, slowly cooling off near the end on long scrolls. Because metal moves more easily when hot, if there are any sharp heat changes, the metal will suddenly bend differently at that point, and your scroll won't look as fluid in its curves. Never have the section you are scrolling end on cool metal if you need to make a long scroll requiring multiple heats.

If you are making repetitive bends or scrolls, make a *jig* to save time. A jig is a duplicate of the scroll, and you bend your scrolls around it when making many pieces that need to look the same. By using a form, you save a lot of the time needed to tweak your bends to make them match.

When using an anvil horn or another mandrel that tapers into a cone to bend or scroll, adjust the angle at which you hold the metal so that it is perpendicular to the section of the cone you are using. When bending on a cone, if you aren't perpendicular to the taper, then one side of the bend will be tighter than the other, creating a lopsided scroll.

FORGING A PENNY SCROLL

1. Begin by setting a section down over a sharper edge on the far side of the anvil with a half-faced blow to create a shoulder to catch on. Once you have a shoulder to catch on, begin to upset the metal by hammering the end back onto the shoulder.

2. Eventually, you will have created a square blob that you can refine into a circle over the far edge of the anvil.

3. Continue to refine the circle by alternating between the edge and face of the anvil. To create a sharp transition where the circle meets the shaft, hammer down and toward yourself.

4. Begin to scroll by hammering it back onto itself as shown.

5. Refine the scroll and tighten it up by hammering down onto the anvil.

6. The finished scroll curves smoothly.

FORGING A SIDE SCROLL

Scrolling flat bar on edge is difficult and slow. To do so, you will need to use your horn to bend it with shearing blows and flatten the bar as it buckles.

FORGING A
FISHTAIL SCROLL

1. To begin a fishtail scroll, taper the end of the rod over the far side of the anvil. Let it widen evenly so that it is symmetrical. Correct any irregularities before it gets too thin because it becomes infinitely more difficult once it is thin and prone to bending.

2. Once it is widened symmetrically, use a small-diameter fuller to incise lines in the taper that follow the outer edges of the flare.

3. Begin the scroll over the far edge of the anvil using shearing blows.

4. Roll the scroll by hammering it back toward yourself as shown. Notice how the heat in my taper is long?

5. Continue to hammer the scroll back toward yourself, adjusting your hammer angle to control the tightness of the scroll. Here, I am hammering slightly downward to tighten the scroll up.

6. Tighten the scroll as needed by rotating the bar and hammering down into the anvil as shown. Repeat the last three steps until you have a smooth scroll of the length you need.

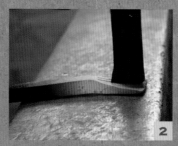

Did you think you wouldn't need math after graduation? To efficiently forge iron and make sure it fits properly at the end, you need to be able to estimate how much growth and shrinkage to expect as you forge. The good news is that iron has a set volume, so you can estimate how it will change with simple geometry. If you are ever in doubt, draw out the project, either actual size or scaled, and measure it with string or a measuring wheel.

CALCULATING STOCK REQUIRED FOR DRAWING OUT:

The formula you use to determine how much parent stock you need when drawing out depends on whether the starting or finished dimensions are round, square, or rectangular.

Because the starting volume equals the finished volume, regardless of its starting or finished shape, you simply use the appropriate formula on either side of an equals sign (=). Most often, you know the final dimensions because you have the project figured out, and you know the starting stock dimensions except the length you require of the parent material. If you have forgotten your high school arithmetic, it's time to brush up! As you progress into making projects such as gates and railings, which have pieces that must fit together precisely, math will come in handy.

To calculate the volume of a rod: Volume = πr^2, or π (pi) multiplied by the radius squared, multiplied by height or length of the metal.

π = approximately 3.14

To calculate the volume of a square or rectangular bar: Volume = length x width x height

CALCULATING STOCK REQUIRED FOR TAPERING:

The math for full round tapers is slightly different from the one for square stock.

To calculate the volume of a round taper: $\pi r^2 \frac{h}{3}$ or 3.14 multiplied by the radius squared multiplied by $1/3$ of the height.

To calculate the volume of a square or rectangular taper: $1/3$ (lwh), or length multiplied by width multiplied by height, and then the total divided by 3.

CALCULATING STOCK REQUIRED FOR BENDS:

While a sharp 90-degree corner doesn't need any extra calculations, you do need to figure out circumferences and arcs when it comes to rings, U bends, and scrolls. Rings are simple: just calculate the circumference. U bends, such as what you would use to make a chain link, can be broken into a circle split in half with two straight pieces stuck between the curves. Simply find the circumference of the two half circles and add in the straight sections.

To calculate the circumference of a circle: $2\pi r$, or 2 multiplied by 3.14 multiplied by the radius of the circle.

5

ADVANCED
BLACKSMITHING
TECHNIQUES

N

ow we that we've covered the more basic blacksmithing techniques, we can move on to some of the more advanced techniques.

FORGE WELDING

Forge welding is the most mystical part of blacksmithing—there is just something about those spraying sparks that grabs the attention of aspiring blacksmiths as well as everyone who sees a blacksmith at work. Forge welding is also one of the most common frustrations for novice blacksmiths who are trying to progress in the craft. One of the most common reasons that a beginner blacksmith has trouble with forge welding is that he or she is not getting the iron hot enough to weld. Familiarize yourself with the different temperature colors and make sure that your forge can burn steel of the size with which you are working. If it can't, you need to look at your fire control or fuel source; rarely do problems occur because of a lack of air supply, although I have known a few mice who've decided to nest in a blower duct.

Forge welding is difficult to explain and also difficult to try the first time, but once you get it, you'll wonder why it took so long to figure out. While, as I mentioned, one of the main obstacles in forge welding is that blacksmiths tend to underheat the metal, it's just as common for a blacksmith to get overzealous and burn the metal. To forge weld, you need to heat the metal until it is almost molten and fuse the pieces by hammering them together. The metal needs to be clean—not oxidized—because scale prevents welding.

TIPS FOR EASIER FORGE WELDS

- Practice the steps and your technique before actually welding. You must be quick to maintain enough heat to weld once you've removed the metal from the fire.

- Don't overdo the air supply because oxygen causes oxides that prevent welds.

- Keep your fire deep to maximize heat and reduce contact with oxide-forming air.

- Once the metal is close to welding heat, back off on the air supply and let the metal soak to ensure that it reaches the proper heat all the way through, not just at the scarf.

- Preheat your anvil if it is cold outside or if you are working on smaller pieces.

- Move your anvil closer to the fire or have a small anvil on your forge for very small pieces.

SCARFING

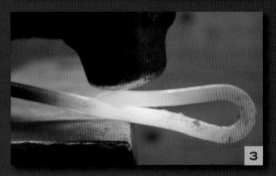

1. If you are scarfing the end of a rod, it is easier to upset the end by making a short bend over the far side of the anvil and forge the end back into itself.

2. Taper the edge over the far edge of the anvil to create a scarf.

3. Welding temperature is bright yellow. In this image, you can see the scarf and how it minimizes the amount of steel that is perpendicular to the welding plane.

For proper heating, you need a good firebed that is free of *clinker*. Clinker robs heat from the fire, forcing you to add more air to get the fire hot enough, which increases the amount of scaling in the weld by increasing the height of the oxidizing layer of the forge. That being said, you can weld in a dirty fire with practice and proper fire control.

Because you need to force the metal together when forge welding, and because you will experience some iron loss from the high heat, you must upset the weld area to account for these losses. For a clean-looking weld, taper the edges so that they blend in seamlessly. The process of upsetting and tapering the edges is known as *scarfing*. A proper scarf is convex so that the flux and scale don't get trapped and prevent the weld. The scarf is also important to minimize the amount of metal that ends up in a shearing plane that won't forge weld together. Forge welding fuses only parallel, not perpendicular, planes, so the edges need to taper to a sharp point (this sharp point is what is called the "scarf"); otherwise, you end up with an unwelded cold shut.

Once your metal is red hot, wire-brush any scale off and flux the weld to prevent new scale from forming as you bring the metal up to welding heat. Make sure that your fire is deep and hot so that you can keep oxygen from getting to the metal. Slowly bring the steel up to a bright yellow—almost white—color so that the surface appears to soften, like butter just before melting.

If you are working with a larger piece of metal, cut the air back and let the piece soak in the heat to ensure that the whole piece of metal reaches welding heat. If just the surface is hot, the center will rob the surface of heat, preventing a weld. To help keep the pieces at welding heat, you must bring more than just the weld area up to welding heat to create a heat sink that will keep the scarf from cooling too quickly.

PUNCHING AND DRIFTING HOLES

There will be many times when you need to either join things together by rivets or make holes for a handle, screw, or bolts. You can drill the holes, but you first must make sure that the metal is wide enough to have strong sidewalls. Punching and slitting holes is often easier than drilling and can maintain the amount of sidewall you need for strength around the hole.

A punch can be round, flattened, or anywhere in between, depending on the shape you want the hole to be. The tip can be flat, as in a standard punch; pointed for a center punch; or rounded off for decorative holes, as you will see in the various projects in this book.

DROPPED-TONGS WELDING

If you work alone, you will need to learn how to forge weld separate pieces by quickly dropping the tongs in your hammer hand so that you can grab the hammer and weld. This takes a lot of practice and prep to get right. Before starting, practice the motions cold and have your hammer on the anvil, ready to grab.

To weld with the dropped-tongs method, prepare the metal the same as you would with any other weld and make sure that you hold the smaller piece in your hammer hand. Quickly move to the anvil so that the pieces are across the anvil face and ready to overlap the weld. The piece in your tong hand needs to be above the piece in your hammer hand to pin it on the anvil and free up your hammer hand to hammer. In one motion, pin the piece in your hammer hand with the piece in your tong hand, drop the tongs in your hammer hand, grab your hammer, and begin to weld. You need to complete the whole process quickly and efficiently because once the piece is out of the fire, you're losing heat.

To punch a hole, begin by relatively lightly hitting the punch to set your mark. Do it lightly in case your mark ends up in the wrong place and you need to move it. If you hammer an incorrect mark too deeply, it will be hard to hide. Once you have your mark, proceed to heavy hammer blows and remove the punch to cool it after two or three hits. If you let the punch heat up too much, it will lose its temper and be more prone to riveting over. Trust me, you don't want your punch riveting into a blind-ended hole!

For smaller metal, you can simply punch through from one side until you feel it get solid as the punch compresses the metal against the anvil face. Once you feel that, flip the metal over and punch the slug out over the pritchel or hardy hole, depending on the size of your punch. To find your mark, you should be able to see a cooler circle where the metal is the thinnest. In thicker metal in which you don't want a large taper from the punch, alternate punching from both sides equally so that the slug ends up just past the middle on one side.

1. Begin by punching through until the punch feels solid when you hammer on it. Cool the punch every couple of hits to prevent it from losing its temper.

2. Flip the metal over; locate the cool spot, which indicates the hole; and punch the slug out over the pritchel hole.

3. The finished hole. In this case, I wanted a taper, so I left the hole as the punch made it.

4. If you have metal in which you want to make a large hole without losing sidewall strength, use an elongated flat punch or a chisel to make a long, narrow hole that you can then open up and make round to keep as much metal in the sides of the hole as possible.

5

6

7

DID YOU KNOW?

When you are punching the slug out, do so when the hole is a reddish/dull orange color. If the metal is too hot, the slug won't shear out cleanly, leaving ragged edges in the hole.

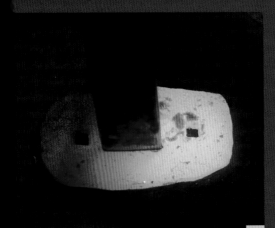

8a

8b

5. A slitting chisel, as shown, is a sharp, pointed chisel that you can use instead of a slot punch to create an elongated hole.

6. Using a drift, which is a punch with a flattened tip to match the slit width, open up the hole and make it round.

7. We will be using a *hammer mandrel*, which is a large drift, later in the book to make the eye for a hammer handle, as shown.

8. Use two parallel slits (a) to create a strip that you can then drift into a loop (b).

   When punching and drifting, if you notice that one side is moving faster than the other and your hole is getting out of alignment, heat the slower side up more or quench the faster side. This makes the side that is slower more malleable so that it catches up to the other side.

CREATING A RIVET

Begin upsetting 1½–2 times the diameter in the vise by hammering lightly and straight down. I find that rotating around the vise and spinning the stock in the vise with each heat corrects any angle I may have in my hammer stroke and thus keeps the upset from folding over. Once you have the head far enough along that there is a shoulder, proceed to assembling the parts and riveting the other side. I never finish my rivet heads in the vise because they will be deformed slightly when riveting the other side.

RIVETING

Rivets used to be the way to fasten most items before welders became common. Rivets are also often used on tools requiring hinges, such as tongs or scissors. Getting a properly upset head is difficult and requires practice. To create a rivet, either forge down thicker stock to create a shoulder for the rivet header to catch when forming the head or use a rod with the same diameter as the hole. If you use the same diameter as the hole, you need to either have a spacer underneath to protect the extra shaft that will make up the other head, or you can begin in the vise. Most often, I end up simply starting the rivet in the vise and finishing the first head after I start the second head. If you do want to use a spacer, it is simply bar stock that is the same thickness as the rivet head material with a hole drilled in it that is the same diameter as the rivet shaft. Creating a rivet is upsetting, so you should heat only up to two times the diameter to make the rivet head; if you heat any more than that, the head will fold over and look ugly.

DID YOU KNOW?

A tenon can be round, square, or flattened, depending on whether it needs to twist or is simply for aesthetics.

Tenons

A tenon is essentially a rivet shaft on the end of a bar, with the bar making up one rivet head. It is often used to join parts together in gates and furniture. To create a tenon, you forge a sharp transition from the parent stock and then draw out the tenon; a spring fuller is a handy tool for this process. After drawing out the tenon, square up the shoulder so that the joint looks tight when finished. You can use a riveting plate and hammer the end down into it, or you can use a monkey tool.

MONKEY TOOL

A monkey tool is a rod with a hole drilled into the end that is the same diameter as the tenon. You hammer the monkey tool onto the shoulder of the tenon to square its shoulder. To make it easier to get a tight fit, grind the end into a slight point. This leaves a depression in the middle, and the edges will spread and fit more tightly once the rivet is finished.

MAKING A TENON, STEP BY STEP

1. To make a tenon, begin by fullering down at the distance you need. Calculate how long the tenon will be once forged to the diameter you want.

2. Over the near side of the anvil, to protect the shoulder you created with the fuller, begin drawing out the tenon. Remember to always go square, and then octagon, and then round if you are making a round tenon.

3. Figure out how much tenon you need to span the pieces it will be riveting as well as the mass for a rivet head ($1\frac{1}{2}$–2 times the diameter, remember?). Proceed to cut it off. In this case, I used my hardy, but you can use a hacksaw or cut-off saw if it is cool when you measure and mark.

4. Square up the shoulders in either the piece itself, a plate with a hole of the same size diameter drilled in it, or a monkey tool. In this case, I am matching the pieces together by marking a side to keep it lined up consistently while I upset the tenon.

5. After heating the tenon and part of the shaft to keep the tenon warm longer, clamp the tenon in the vise and assemble the pieces to begin riveting. As you can see, it has begun to fold over slightly. To correct this, simply switch sides to hammer it back into alignment.

6. And there you have it, a finished tenon!

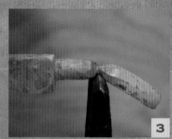

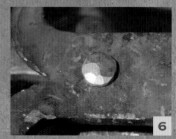

COLLARS AND WRAPS

Another way to join pieces together is by collaring or wrapping. You can also use a collar to add mass to an area by forge-welding the collar to the specific area. This is quite common for finials on a gate, where it is impractical to upset enough mass to forge that ball or large leaf. Collars can wrap around both of the pieces you wish to join or pass through a hole punched in one of the bars, which is quite common for screens or sign brackets.

There are two types of collars: overlapping and butt. An overlapping collar has two tapered ends that overlap on one side. A butt collar has the ends butting together without any tapering. Other than tapering the ends and using slightly different calculations to figure out how much length you need, they are made in the same way. Of course, you can also experiment with wire wraps. I will be showing you a method that requires no special tooling because you probably want to get to blacksmithing right away. If you are doing many collars, however, making a jig to quickly and consistently fold them will make your life easier.

A forge-welded collar can be used to add mass to the end, as in this toilet paper holder.

OVERLAPPING COLLAR

To make an overlapping collar, begin by tapering both ends of your collar material (**1**). Make sure that the tapers are identical so that they overlap correctly. Following tapering, bend one leg 90 degrees in the vise. How long each side is depends on the pieces it needs to join. Next, using high heat and a piece of metal with similar dimensions clamped in the vise, pin the first bend against the rod with a pair of tongs and proceed to wrap the collar around the bar until it is completely wrapped around (**2**). Finish by hammering the collar tighter on the anvil.

MAKING YOUR OWN TOOLS

The great thing about having a blacksmith forge is that you can make your own tools or refurbish old tools very inexpensively. Historically, journeymen blacksmiths had to make their own tools before setting off on their own and making tools for everyone else. The tools in this chapter will get you started forging and will help out around the house and farm.

MAKING YOUR FORGE TOOLS

Your first project should be a set of fire-maintenance tools because without proper fire maintenance, you will be forever inefficient as a blacksmith—dirty fires will prevent you from forge-welding, clinker will rob your fire's heat, and you will burn more fuel than you need to as the fire constantly grows more inefficient and larger. Forging a poker, rake, shovel, ladle, and coal scoop or scuttle will help you work on tapering, rounding, dishing, and riveting.

MAKING A MATCH

When you are making anything that needs to match, always start with stock of equal dimensions. If you have unmatched types of steel, I find it easiest to forge them into matching stock before performing the next step. For example, if I have differently shaped (e.g., square and round) or sized pieces of steel, I will forge the larger piece to the same dimensions as the smaller piece and then taper both pieces to match. You only need to forge the piece to match as far as the next step, and you need not change the dimensions of the whole piece of metal unless you figure out the matching volumes and cut off that exact amount. I always prefer to err on the side of caution, so I like there to be extra metal to cut off at the end. You can always find small one-off projects to use up the cut-off pieces.

THE UBIQUITOUS S-HOOK

Every blacksmithing how-to book includes a section on forging an S-hook as a beginner's project, and I would be amiss if I didn't as well. Besides, S-hooks are useful for hanging up your forge tools as well as for other things around the house and farm. The hooks can be plain, or they can be as decorative as you want, with fancy scrolled ends and twists.

1. Begin by tapering one end of a ¼-inch round rod over the far edge of the anvil until the taper is approximately ¾ inch long. Cut off the rod at 5½ inches and repeat tapering the other end.

2. Scroll the ends of the taper by beginning to bend the tip over the far edge of the anvil and then rolling it back as shown.

3. Quench the scroll in water to protect it and then bend each of the ends around the horn of your anvil or a ¾-inch-diameter rod. If you are making multiples, a rod will make them more consistent.

4. Finish the bend by hammering straight down to straighten the tip of the bend and create a deeper hook.

FIRE POKER

STARTING STOCK
- ³⁄₈-inch rod, 2 ½–3 feet long

TOOLS REQUIRED
- Hammer
- Anvil

This is the easiest project in this book other than the S-hook; you need to simply taper and round one end of a rod and then make the handle. Don't let the simplicity of this project fool you, though: a poker is an important tool, especially if you use coal, because the fire will settle and block your air supply. Every so often, it is good to lift and break the coal up with your poker to let more air into the fire. The poker will also let you lift any clinker out of the fire and blow the ash out of the way.

1. Begin by tapering one end of the rod over the far side of the anvil until the taper is 5 inches long. I always taper over the far side of the anvil, if I can, to protect the anvil face from the edge of my hammer. As you can see in the photo, one corner of the hammer would hit the anvil face before I hit the rod near the tip of the taper. This leads to dings in your anvil face as well as marring the smoothness of your finished pieces because those dings will be pressed into the metal you are forging. The only way to achieve a perfectly smooth forged finish is with smooth hammer and anvil faces.

2. Once you have a 5-inch-long gentle taper, proceed to round the taper, remembering to go from square to octagon to round as previously described and as shown here. Hold the adjacent corners perpendicular to the anvil face and hammer straight down. If you don't follow the proper steps, you will end up chasing a corner around, forming cracks near the tip.

3. On the handle end, square up ¾ of an inch. Once it is square, bend it 90 degrees over the far side of the anvil.

4. Approximate your hand width, with an extra inch for comfort plus 2½ inches for the bend, from the end. In my case, the total length was 7 inches, and I have average-sized hands. Mark that length on the anvil and, once the rod is heated, line the end of the rod up with the chalk mark. Next, lift the end up and start bending the rod on the edge of the anvil. Once you have your mark, bend it over a round object, such as an anvil horn or a piece of pipe or rod in the vise. Just make sure that when you bend the handle around, you bend it so that the tab you bent over at the end bends around to touch the shaft of the poker. If you do end up bending it the wrong way, simply twist the tab so it is correct, and no one will know the difference.

Close the handle over the anvil face. As you do this, a bend will form in the bottom of the handle, as you will see in the final piece.

Congratulations! You've finished your first blacksmithing project, and now you can say that you're a blacksmith!

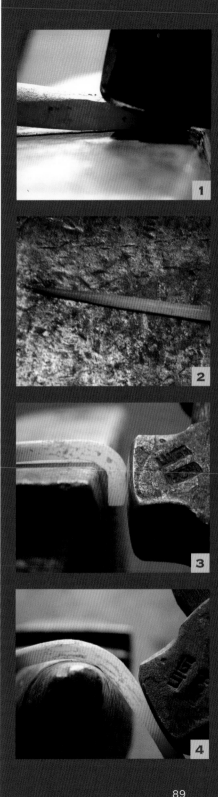

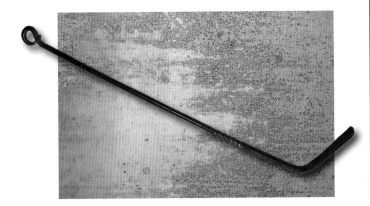

FIRE RAKE

STARTING STOCK
· ⅜-inch rod, 2–3 feet long

TOOLS REQUIRED
• Hammer
• Anvil

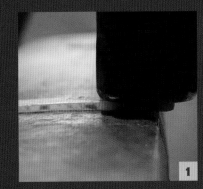

1. With full-face hammer blows, flatten 5 inches of the rod down to ⅛ of an inch thick on the face of the anvil, as shown. Keep your hammer and rod parallel to the anvil face and allow the metal to spread wide; this will become the rake that you will use to move coal in and out of your fire. If you need more width, remember that you can use a fuller to move the metal in one direction. In this case, to make the piece wider, you need the curve of the fuller to be perpendicular to the length of the rod. This pinches the metal side to side rather than lengthwise, much like you would use your fingers to spread clay. In other words, your peen needs to be parallel to the length of the rod.

2. You want to end up with a shape similar to that shown in the photo of the rake's end. Offset the widening by placing the metal on edge across the hammer face and hammering it at the point where the metal widens and anywhere else where it is not lying flat on the anvil. Of course, success relies on a straight anvil face, so you should double-check the straightness of the rake with a straightedge or ruler.

CONTINUED

3

5

7

4

6

8

3. Bend the rod 90 degrees over the far side of the anvil, just before the flattened portion, with shearing hammer blows. Square up the bend on the anvil corner as shown.

4. To add strength to the bend you just made, taper the outside of the corner. This widens the corner and puts more metal in line with the direction in which the corner will want to straighten, thus strengthening it. You don't need to do this, but it will help keep your rake from straightening with use. Taper the corner to ⅛ inch at the far edge to prevent damaging your anvil.

5. For the rake handle, we're going to use a simple wrapped taper eye to distinguish it from the other forge tools. Taper and round the end, following the recommended procedures in the poker project, so that you have 5 inches of rounded taper.

6. Bend the start of the taper slightly further than 45 degrees over the far edge of the anvil and then bend the tail back around the horn or rod until it looks like this image. The interior diameter of the ring should be ¾ of an inch.

7. Using your tongs or a pair of pliers, wrap the tail around the shaft of the rake as shown. Make sure that you tuck the tail in tight to prevent cutting yourself on the sharp tip.

8. This time, let's add a twist! To twist round metal, you need to first square the rod; you can then twist it by clamping the rake in the vise where you want the twist to start. Because you made a round handle, simply heat only the section you want twisted, put a rod through the handle, and twist while pulling straight (see next paragraph). If you heat a larger section than you would like, quench the unwanted part.

    Because this is your first twist, don't worry too much about the length of the twist; just concentrate on keeping the twist straight and even. A trick to keeping a twist straight is to pull straight backward consistently while you turn. This will help keep you from bending the rod while twisting it. Use a wooden mallet to straighten the twist.

TELL THEM APART

Keep track of your fire tools by using a differently shaped handle on each of them. Often, you will be grabbing for them while watching the fire, or they will be mixed up in a pile of tongs or steel on your forge table.

FIRE SHOVEL

STARTING STOCK

- Handle: ½-inch rebar rod, 17 inches long
- Shovel: 18-gauge sheet metal, 3 inches x 4 inches
- Rivets: ¼ inch rod, at least 3 inches long

TOOLS REQUIRED

- Cross-peen
- Hammer
- Anvil
- Method of "dishing" sheet metal
  (e.g., swage or wood stump)
- Method of cutting sheet metal
  (angle grinder with cutting disc, cold chisel, hacksaw, bandsaw, oxy/acetylene torch, plasma cutter)
- Hot cut or chisel
- Punch
- Vise
- Drill and ¼-inch metal drill bit

There are a few different ways that you can make a shovel. In this project, you will cut the shovel out of a piece of sheet metal and rivet it to the handle. Because I like to salvage to both reduce my costs and help the environment, for this project I used an old rusted-out garden spade, but you can use any similar thin-gauge metal. I wouldn't go any thicker than ⅛ of an inch for weight, though.

1. Cut out the shovel blank, 3 x 4 inches rectangular, and round the corners with a grinder or file. In a V-swage, use your cross-peen to dish the shovel, as shown. Using a cupped dishing tool or a block of wood and a ball-peen, dish one end of the shovel into a cup.

2. The shovel blank should look like this when it is finished and ready for the rivet holes.

3. Over the near side of the anvil, flatten the end of the rebar and let it spread as much as possible. Use your cross-peen to spread the metal wider, if needed; you want the metal to be wide enough to keep the shovel from twisting once riveted. Center-punch two holes down the center of the widened end, ¾ inch apart. Center punching will create a depression to seat the drill bit, preventing it from skating around and missing your mark.

4. Next, taper behind the widening slightly to help balance the weight of the shovel. Keep the widened tab over the far edge of the anvil; remember to go from square to octagon to round. As you get closer to the tab, use a ball-peen or rounding hammer on the horn to clean up the transition where it begins to widen.

5. Using a rounding or ball-peen hammer and a curved shape, such as the anvil horn, taper and round 1½ inches of the handle end.

6. Flatten 5 inches of the handle slightly and put a small curve into the handle over the horn. Proceed to bend the tip you drew out 45 degrees, and bend it around with a pair of scrolling tongs to create a loop to hang the shovel. Clamp the shovel to the handle to drill the holes and then rivet with ¼-inch rod. Bend the handle near the shovel to a comfortable angle for you to use.

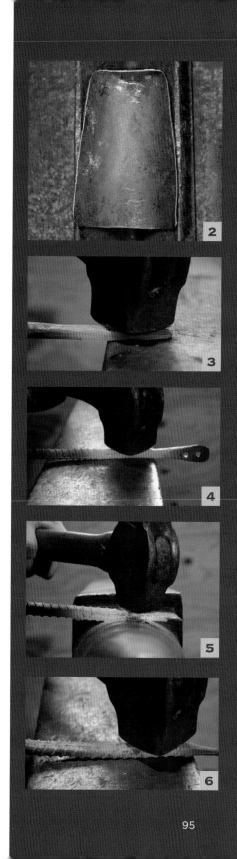

FORGE LADLE

STARTING STOCK

- Handle: ⅜-inch square rod, 2 feet long
- Ladle: 18-gauge sheet metal, 6-inch-diameter circle
- Rivets: ¼-inch round mild steel

TOOLS REQUIRED

- Hammer
- Anvil
- Method of "dishing" sheet metal (e.g., swage or wood stump)
- Method of cutting sheet metal (angle grinder with a cutting disc, cold chisel, hacksaw, bandsaw, oxy/acetylene torch, plasma cutter)
- Hot cut, chisel, or hacksaw for partially cutting off rivet
- ¼-inch punch
- Vise
- Drill and ¼-inch metal drill bit

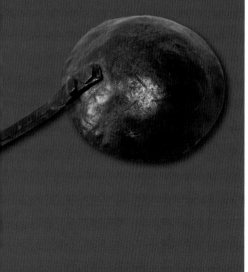

1. Mark a 6-inch diameter circle out of your 18-gauge sheet metal. In this case, I cut it out of the 55-gallon drum lid that I also used for the coal scoop (see page 99). Using your preferred cutting method, cut close to the line and then file or grind the blank to the line you drew. This will give you a nice, round ladle blank.

2. I used a cupping swage designed for forging spoons and ladles; you can buy or make a similar one. You also could use any similarly dished shape, such as the metal cap from an old oxygen tank or a wooden stump stood on end.

3. With a rounding hammer or larger ball-peen hammer, begin dishing from the center of the circle and continue working your way out in a spiral pattern. Work in overlapping hammer blows to smooth the ladle.

4. To start the handle, taper one end of the ⅜-inch square rod on the far side of the anvil with full-face hammer blows until the taper is 1½ inches long. Let it widen (this will later become a fishtail scroll).

5. Curl the end of the handle taper over the rounded far edge of the anvil and begin scrolling by holding the rod flat on the anvil and continuing to hammer back toward yourself. Remember that consistent heat makes your scrolls smoother. As you notice irregularities, correct the scroll as shown by lifting the rod off the anvil and hammering. If your scroll doesn't end up smooth and symmetrical, unroll it with a pair of scrolling tongs or pliers and begin again. You can correct some minor alignment issues by hammering it on edge.

6. Bend 6½ inches approximately 45 degrees to one side over the far side of the anvil and then bend 5 inches around your horn, mandrel, or

CONTINUED

¾-inch rod held in a vise. The diameter of the ring should be ¾ of an inch.

7. Line up the bend of the ring so that 1 inch of the end taper is flat against the shaft, as shown. To flatten it, place the loop over the far edge of the hole with the scroll up and use your cross-peen to get between the ring and the scroll without damaging them.

8. Quench the handle end and heat up the other end to create the area to attach the ladle. Flatten 1½ inches over the near side of the anvil down to ⅛ inch thick. Create a step by using a half-faced blow, as shown. Let the steel widen to create as much contact area as possible for the ladle. After flattening the end, punch two ¼-inch holes in the flattened portion, ½ inch in from each end. Once you've punched the rivet holes, curve the end over the horn so that it matches the curve of your ladle.

9. Mock assemble the ladle and handle, with the handle on the outside, and mark where the punched holes line up on the ladle bowl. Drill two ¼-inch holes in the ladle at your marks. Make two ¼-inch-diameter rivets to attach the handle and then rivet the ladle to the handle. I find it easier to assemble the ladle with the preformed head on the outside and rivet the inside head while supporting it on the near edge of the anvil, as shown. You can rivet these cold, but I prefer to heat them to red in the forge before fully cutting the prepared rivet off its parent stock. If you don't feel comfortable making your own rivets yet, purchase them from a local welding or agricultural store.

Note: Remember that the shaft length of a rivet with one preformed head should be the total width through which the rivet will be passing plus two times the shaft diameter—in this case, a maximum of ½ inch extra on a ¼-inch-diameter rivet.

COAL SCOOP OR SCUTTLE

STARTING STOCK

- Handle: $1/2$-inch round rebar, 25 inches long
- Shovel: 18-gauge sheet metal, $7^3/_4$ x $9^1/_2$ inches
- Rivets: $1/4$-inch rod

TOOLS REQUIRED

- Hammer
- Anvil
- Method of "dishing" sheet metal (e.g., swage or wood stump)
- Method of cutting sheet metal (angle grinder with a cutting disc, cold chisel, hacksaw, bandsaw, oxy/ acetylene torch, plasma cutter)
- $1/4$-inch drill bit and drill

1. As I mentioned in the ladle project, I used the same metal drum lid for the ladle and for this project; shown is the shovel blank for the coal scoop. The blank is 7¾ inches wide by 9½ inches long, and the inside chalk lines, where I will be bending the sides up, are 1 inch inside. Clean up any sharp edges after cutting out the blank.

CONTINUED

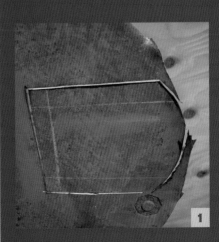

2. Notch the end of the blank as shown, in line with the inside bend lines. Because the sides will be 90-degree corners, the notches will make it easier to dish and fold the edges up without wrinkling.

3. In the vise, fold both edges 90 degrees along the 1-inch chalk line. Using a dishing stump or swage and a rounding or ball-peen hammer, dish the end up, as shown. Once it is dished up, fold the tabs behind to help strengthen the shovel.

4. On the face of the anvil, flatten 7½ inches of the rod and widen it to a minimum of ¾ inches. You will split this into two later, so you need as much width as you can get. You can use a cross-peen to help widen it once you have begun to flatten it.

5. Flatten any irregularities in the edges without drawing out the width (remember that you need all of the width you can get).

6. Set down the end of the widened area to create a slight shoulder for the edge of the shovel to sit in, but no more than ⅛ inch. Neck down and slightly taper the area just behind the widening, as you did for the forge shovel.

7. With your cutting tool, split 6¼ inches of the widened area in half. Over the near edge of the anvil, using half-faced blows, set down ½ inch on the same side of both ends of the slit material. These will be drilled and riveted to the sides of the scoop.

CONTINUED

CHISELS AND PUNCHES

Chisels, called "sets" if they are handled, can be intended for either cold work or hot work. The only difference between cold and hot chisels is the angle of the cutting bevel and how much metal there is behind the edge. Chisels used for cold work need a 60-degree angle on the edge, whereas a hot chisel can be much slimmer because hot metal cuts more easily.

Handheld chisels are very easy to forge, requiring only a taper, but the difficult part of making a chisel is heat-treating the steel to make the cutting edge hard enough to be resilient to dulling but not so brittle that it chips. Generally, you want to temper the steel to a brown oxidation color.

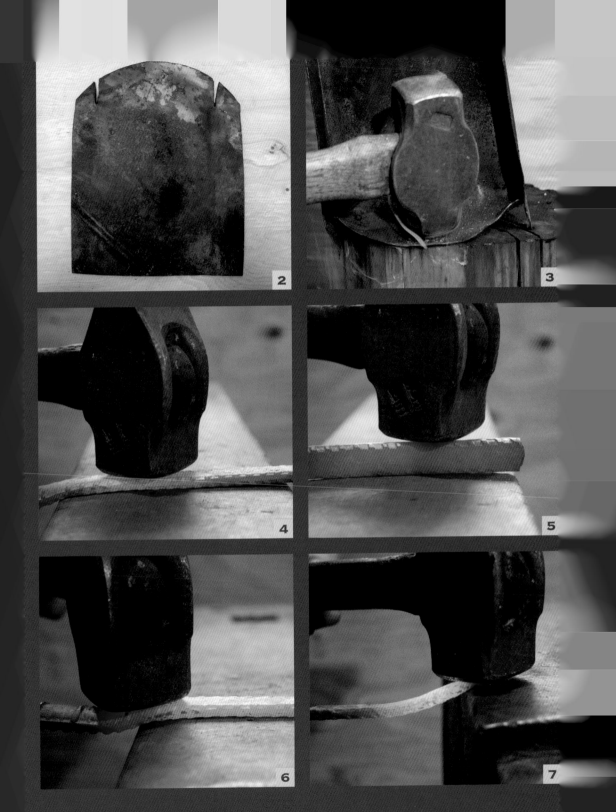

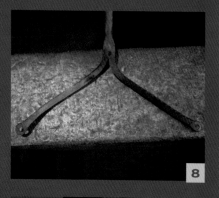

8. Center-punch and drill ¼-inch holes in the center widening that was left intact and in the two split ends. Bend the ends, as shown, to match the outside of the coal scoop.

9. Set down adjacent 90-degree sides over the near side of the anvil with half-faced blows. Taper it into a 3-inch-long point.

10. This is what the end should look like after setting down the sides. Continue tapering the end to a sharp point.

11. With your scrolling tongs, create a scroll and proceed to bend the handle around the horn, as shown. Stop once the tip hits the shaft of the handle. Heat the corner just before the hammer (refer to the photo) and bend the handle flat over the anvil face.

DID YOU KNOW?

Two hexagonal nuts from your local hardware store held with two flat sides touching will give you a 60-degree angle. This is an easy way to ensure that your cold chisels are ground properly.

One of the best chisels you can have for removing rivets has a curved edge and a single bevel.

HANDHELD CHISEL

STARTING STOCK

- ³/₄-inch medium- to high-carbon steel 10–12 inches long

TOOLS REQUIRED

- Hammer
- Anvil
- Appropriate quenching medium: oil for high-carbon steel or water for medium-carbon steel

1. To forge a straight chisel, begin by tapering two sides.

2. Correct any widening so that the chisel edges are in line with the shaft edges. Now heat the chisel up to nonmagnetic and let it cool by the fire. Once cool, heat half of the chisel up to nonmagnetic and quench one-third of the tip in water or oil. Quickly polish the tip and watch the colors run as the remaining heat tempers the tip. Quench it again when it reaches straw yellow at the tip.

   Sharpen the chisel to a 60-degree edge, taking care to not heat the tip up with the grinder, and test to see if your chisel will cut cold metal. If the edge dulls right away, it needs to be tempered to a harder temper. If the edge chips, it needs to be a softer temper.

   All chisels are made the same way, but with a different edge ground in. For example, a slitting chisel has a V-shaped point ground into the edge rather than a straight edge. To add a handle, punch and drift an eye for the handle, as I will demonstrate in the next project, the handled slot punch.

HANDLED SLOT PUNCH

STARTING STOCK
- 1¼-inch medium- to high-carbon steel, 6 inches long

TOOLS REQUIRED
- Hammer
- Anvil
- Handheld slot punch

1. To make a handheld slot punch, simply forge a longer tapered chisel (see previous project) and leave the end flat. Relieve any sharp corners around the edges of the bevel to prevent any sharp corners in the hole when you later drift it open.

2. For this handled slot punch, I used the male joint on a piece of oilfield sucker rod. The transition to ¾-inch round from larger 1¼-inch steel makes for an easier time forging. You can use any 1¼-inch steel, but you have to forge the punch end down until it looks like that shown in the photo. Flatten the taper on a 45-degree angle to where the handle will go through to make it easier for you to line up the punch while holding the metal.

3. Using your handheld slot punch, punch through the flattened area from both sides so that the slug meets in the middle. This will minimize the taper of the eye, making drifting easier.

   After you have punched through the eye, continue to drift out the eye using your slot punch; enlarge it by hammering the sides into an eye mandrel to draw them out. Work from both sides to minimize the amount of taper. You can make a separate mandrel to protect your slot punch if you wish, but I just reforge and heat-treat my slot punches as needed.

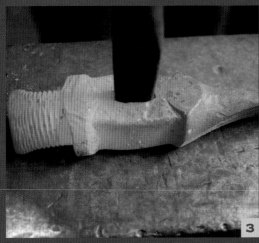

PUNCHES

Punches are made similarly to chisels except you forge the tip into a point rather than a two-sided taper. The tip of a punch can be left square, rounded, or flattened. The main difference between a chisel and a punch is the edge: chisels have sharpened edges for cutting, while punches are flattened to compress and shear slugs out of metal.

EYE DRIFT AND MANDREL

STARTING STOCK

- 1¼-inch axle shaft or similar medium-carbon steel, 8 inches long

TOOLS REQUIRED

- Hammer (the heaviest that you can swing comfortably) or striker with sledge
- Anvil

1. Begin by tapering the drift to 5 inches long on two sides. Correct the widening and taper the other two sides in so that the end of the drift matches your slot punch, ⅞ x ⅛ inch in this instance. Flatten the portion behind the working taper to 1½ x 1 inch for 3 inches and then taper the hammer end slightly. This reverse taper lets you hammer the drift all the way through, rather than trying to reverse it out of the hole. Because they are not doing any cutting or hard labor, I don't bother to heat-treat my drifts because they lose their temper once in the hot hammer eye. Simply normalize or anneal the piece at the end of forging to reduce any grain stresses.

2. Always correct the fish lips that develop as you forge this thicker steel. Because your hammer blow is strongest at the edge, the outside of the metal moves the most. If these fish lips continue to develop, they will be tapered down into two ¹⁄₁₆-inch thin pieces of metal with a cold shut between them. It wouldn't take much to break the tip of the drift if that happens. To correct it, simply upset the end back again until it is flush.
   Note: If you end up with cold shuts at the end of the process, you can forge weld it together to avoid forging another one, but it will be a weaker tool. It is best to prevent such problems from the start.

TONGS

STARTING STOCK

- Two pieces of ³/₄-inch round rod, 16 inches long
- Rivets: ³/₈-inch round rod, more than 3 inches long for each

TOOLS REQUIRED

- Hammer
- Anvil
- Round punch

PERFECT FIT

If your tong rivets are loose, peen them tighter. If they are too tight, heat the metal up and open and close the tongs a few times to loosen the rivet.

1. Begin over the near side of the anvil with half-faced hammer blows. Set 2 inches down to less than half of the starting diameter to allow for finishing later. Draw out the jaw, again leaving a little extra for finishing. The jaws should taper toward the tip for maximum strength.

2. Rotate the stock 90 degrees away from your body and hold the tongs at a 30-degree angle. Set down at the beginning of where you set down in the first step. To prevent thinning the stock down too much before you are finished forging to the proper dimensions, don't forge it down to half of the original diameter.

3. Rotate the stock another 90 degrees away from you and set the rein side of the boss down over the far side of the anvil with a half-faced blow, the same way that you did in the first two steps. The boss for the rivet should be 1¼ inch back to allow for finishing. Make sure that the taper is away from the boss you've created. Proceed to finishing the jaw up and draw out the reins down to ½ of an inch on the end. You can cut off the jaws a few inches past the boss and forge-weld the jaws onto ½-inch round to save drawing out the reins.

4. Using a fuller, create a depression in the middle of the jaws to grab round and square stock better. With a chisel, cut grooves into the sides to help grip flat bar. Punch and drift out a ³/₈-inch hole so that the rivet fits snugly. Repeat on the second blank, just as you did for the first one; they need to be exactly the same.

File the inside of the boss flat so that the two tong blanks fit together nicely. Test-fit with a nut and bolt instead of with a rivet, in case you need to separate the tongs and file more off the inside. Once you are happy with the fit, make a rivet out of ³/₈-inch rod and rivet the two pieces together. Heat the boss up and open and close the tongs a few times to loosen the hinge after riveting.

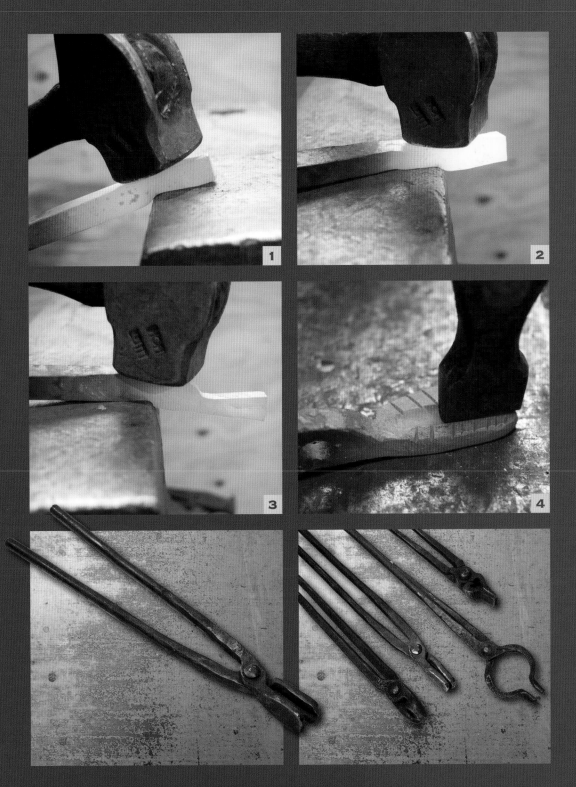

CLAW HAMMER

STARTING STOCK

- 1-inch tie rod or similar medium-carbon steel, 6 inches long (You can increase the diameter of steel if you want a heavier hammer.)

TOOLS REQUIRED

- Hammer
- Anvil
- Slot punch
- Hammer eye drift
- Cold chisel
- Cutting plate

1. Begin by slightly upsetting the hammer end to create a ⅛-inch wider end. It's very slight, but it will help balance the hammer to the front. Flatten 1½ inches from the hammer head end to make punching easier. It should be no less than 1 inch thick once flattened; you just want to help keep the piece from rolling away while you slot-punch.

   Using the slot punch you made previously, punch through from both sides. Set the punch 2 inches from the hammer head and in the midline. Remember to use light blows at the start so that you can move the punch if needed.

2. Once you've punched the hole, begin to drift the hole with the eye drift you just made. Because I fasten the head with a wedge, I like to drift more from the topside to create more taper on the top so that it wedges more securely than a flat-sided eye. Draw out the eyes over the anvil to help stretch the metal rather than having

the drift pull it wider, which can lead to cracks. If the eye is moving more on one side than the other, or getting off center, draw out the shorter side with the mandrel in place and with the other cheek protected in the hardy hole or in your opened vise. This will help keep the eye symmetrical and centered.

3. Over the near side of the anvil, taper down the claw end of the hammer. (*Note:* If this were a cross-peen, you would want a steeper taper.)

4. With a 60-degree-edged chisel, slit the claws from the underside. (*Note:* Although a cutting plate is not pictured in the photo, be sure to use one to protect your anvil face.)

By cutting through from one side, you create a taper to catch the nail heads automatically. If you want to make a leafing hammer, dishing hammer, or cross-peen hammer, simply dress the taper to the shape you want rather than splitting it.

Proceed to heat-treating the hammer head. Anneal and harden as you normally would, but this time, temper the head by heating up your mandrel and fitting it in the eye. Quench when the face turns straw yellow.

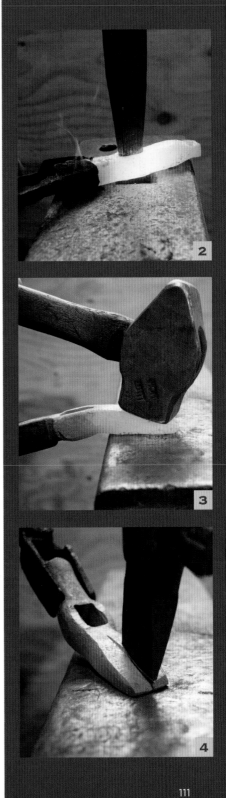

CROWBAR

STARTING STOCK

- 1-inch round or octagonal medium-carbon steel, 18 (or more) inches long

TOOLS REQUIRED

- Hammer
- Anvil
- Chisel
- Cutting plate

I made this crowbar with a shorter shaft simply because I had many longer ones that didn't fit in tight places. You can adjust the length of your starting stock according to the desired length of crowbar.

1. Taper one end of the bar 2 inches and the other end 2½ inches, down to ⅛-inch thickness on both ends. The 2-inch end will be the chisel end, and you will split the 2½-inch end into the claws.

2. While tapering, hold the taper on edge and correct any widening that occurs. You will need a slight taper to the width, as shown.

3. Using a cutting plate to protect your anvil face, slit 1½ inches of the longer taper down the middle. Because you want a taper inside the slit to help grab nails, only slit from one side with a cold chisel, as shown. The bevels of chisel will taper the inside of the slit naturally.

4. In the vise, clean up the end of the slit by holding the chisel at a slight angle so that there is a taper past the end of the slit to house a nail head.

CONTINUED

HEAT-TREATING THE CROWBAR

If you decide to heat-treat the crowbar, heat one end to nonmagnetic again and quickly quench it in water. After quenching, polish the end and slowly reheat until the metal turns blue. Quench the crowbar again and then repeat the steps on the other end of the bar. Personally, I prefer to straighten bent bars than go through the hassle of forging a new crowbar, so I simply normalize by letting each end cool slowly at the side of the fire after heating one last time to nonmagnetic when I'm done forging an end.

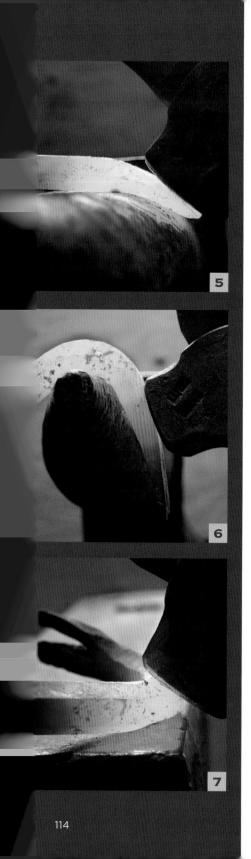

5. Over the horn or a curved piece of steel, bend the chisel end slightly to create a prying surface.

6. Bend the claw end of the crowbar 4 inches from the end, as shown, using your anvil horn or a large rod. Put a slight curve into the prying end before the claws, as shown, to help make prying easier.

7. Taper the bend as you did with the fire rake to strengthen the bend (see Step 3 in the rake project). Again, do this over the far side of the anvil, as shown, to protect the anvil face from errant hammer blows.

    After you are done forging, normalize or anneal the crowbar by heating it to nonmagnetic and either leaving it by the fire or burying it in ash so that it cools slowly. This step removes any grain stresses that occurred during forging. Once it is normalized or annealed, you can either leave it as is or heat-treat it to improve its strength. If you leave it at its softest, it is more likely to bend; however, if you temper it too hard, it will be more likely to break.

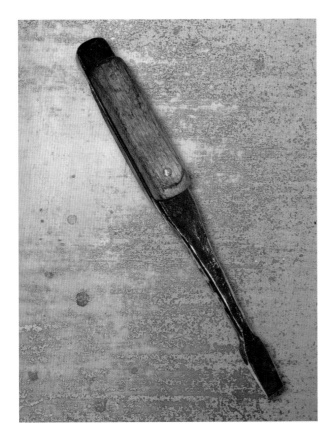

FLAT-BIT SCREWDRIVER

STARTING STOCK
- ³/₄ x ¹/₄ inch rectangular medium-carbon steel, 6 inches long

TOOLS REQUIRED
- Hammer
- Anvil
- Spring fuller or guillotine
- Brazing rod
- Drill and ¹/₈-inch metal drill bit

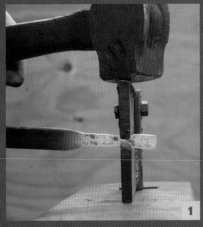

1. Taper the end down to ½ of an inch; the taper should be 4 inches long. Next, fuller ¾ of an inch back from the end to define the transition between the neck and the head of the screwdriver. The end dimensions of the transition to the head will be ¼ inch, so fuller down to ⅜ of an inch to start. You will forge it to the final dimensions later when finishing the screwdriver.

2. This image shows the guillotine fuller being used on a different project. As you can see, there are different-sized spaces ground out so that the guillotine can be used for fullering pipe. You can make a guillotine fuller if you have a welder, a file or grinder, and a

CONTINUED

3a

3b

4

5

drill press. I made this one from mild steel; the bottom jaw is fixed, while the top jaw is hinged on a bolt.

3. Once you've fullered the neck, forge it down to its final dimensions, gradually tapering down to the head transition, which is ¼-inch square. To do this, alternate between the near (a) and far (b) edges of the anvil. Once you've finished the taper, chamfer the edges of the neck by holding the edges perpendicular to the anvil and forging them back into the mass; this is the same technique as you would do to begin rounding the metal. Finish by creating a flat taper on the end of the screwdriver as you would for a chisel. File the end smooth and square so that it matches the screw head it will be used on.

4. Cut the screwdriver off any remaining metal so that the handle is 4 inches long. Either forge or file a slight curve to the end for aesthetics and to get rid of any sharp edges. Drill ⅛-inch-diameter holes in the handle that are the same dimensions as the brazing rod, 2½ inches apart, centered on the handle. After you've drilled the holes, proceed to heat-treating the screwdriver by heating it to nonmagnetic and letting it cool by the side of the fire. Once it is cooled, heat the tip and neck back up to nonmagnetic and then quench only the tip. Next, quickly polish the head of the screwdriver so that you can see the temper colors run. Quench

the screwdriver completely once the tip turns yellow.

5. Cut a piece of hardwood 3 x ¾ x ¾ inches. Round the ends to create the profile seen here and then split it in half to create the two identical halves that will make your screwdriver handle. Match the holes drilled in the handle of the screwdriver body by clamping the wood to the screwdriver and drilling through the pre-existing holes in the metal. Next, countersink the holes on the outside of the wood to house the brass rivet head. Sand the pieces smooth and round the corners. If the handle is too wide, sand the wood slabs until the screwdriver feels comfortable in your hand when it is mock assembled.

   Remove the flux and cut two rivets from ⅛-inch brazing rod, which you can get from most hardware, agriculture, or welding-supply stores. Measure the total width of both handle slabs and the handle of the screwdriver as assembled and add two times the diameter of the brazing rod, ¼ inch in this case. The extra metal will become the rivet heads once they are peened.

   Assemble the handle and, with a ⅛-inch spacer under the handle to protect one side of the rivet, begin to peen the rivet. Once one side is peened, flip to the other side and peen it without a spacer under the handle. Repeat the process on the second rivet, and you have a brand-new screwdriver!

NAIL AND RIVET HEADER

STARTING STOCK

- Header: 1-inch round medium-carbon steel, 1 inch long
- Handle: 1/4-inch round, 2 feet long

TOOLS REQUIRED

- Hammer
- Anvil
- Means of cutting a rod to length
- Square punch (See the Handled Slot Punch project)
- Vise
- Hot-cut hardy

1. Cut off 1 inch of 1-inch diameter round medium-carbon steel. Medium-carbon steel will be more resilient to errant hammer swings than mild steel. In this photo, I am using my anvil hardy, but you can use any cutting tool, such as a hacksaw, cutting torch, or abrasive saw.

2. Upset the header with lighter hammer blows to rivet over both sides and create a lip for the handle to sit in. Flip the piece and work both sides evenly to create a lip on both sides. Form a slight dome on one side to make heading the nail easier.

3. Using a square punch with a 1/8-inch square end (a), punch completely through from the back side (b). Once the punch feels solid when you hit it, shear out the plug over your pritchel hole (c). Only punch and drift the hole to 1/4 inch from the top to create a slight bevel around the hole to prevent damaging the nails you make. Most household nails are made from 1/4-inch nail stock, and you need a slight lip to catch the nail to head it.

   After punching the header, the dome will be flattened and need some dressing. Be careful to keep from damaging the hole you punched. If you do need to repair the hole after dressing the dome, use the punch again; be careful not to punch too deep or the hole will be too large.

4. Center-punch to mark the middle of the 1/4-inch round rod. Heat it and wrap it around the cooled header, which is held in the vise. Cross over the handle ends tightly against the header and begin to twist. Once you've started the twist, clamp the handle in the vise and continue twisting the rods tightly with your tongs, as shown. Straighten the header handle on the anvil, making sure to rotate the handle in the same direction that you twisted. If you rotate opposite to your twist, it will loosen the header, leading to much frustration.

CONTINUED

5. After straightening the rods for the handle, clamp 1½ inches of the end in the vise and twist with the tongs, as shown. Once the ends are twisted together, cut one of the rod ends at the end of the twist. Blend the cut end of the rod into the twist around the other rod and proceed to taper the remaining end.

6. Over the far side of the anvil, square-taper the remaining end with full-faced blows. Continue through the steps of rounding the taper.

7. With your scrolling tongs, or over the far side of the anvil, bend the taper 45 degrees at the end of the handle twist.

8. Curl the taper around with your scrolling pliers, horn, or 1-inch rod that is held in the vise.

9. Once you've finished your header, it is only fitting to make a test nail (see project in Chapter 7) from which to hang the header.

HOLD-DOWN

STARTING STOCK
- ½-inch mild-steel rod, 24 inches long

TOOLS REQUIRED
- Hammer
- Anvil

1. Flatten 1 inch of one end of the rod to a ¼-inch thickness. Proceed to bend the rod around the horn at the center point.

2. Flip the bar so that the untouched side is up and then bend the side with the flattened end over the horn, as shown.

3. To use this tool, tap the straight end into the pritchel and use the foot you created to hold pieces tightly on the anvil while you chisel or fuller them.

SPRING FULLER WITH ADJUSTABLE HARDY SHAFT

STARTING STOCK
- ¼-inch by 1-inch flat bar, 40 inches long

TOOLS REQUIRED
- Hammer
- Anvil
- Vise

1. Begin by bending 8 inches 90 degrees over the far side of the anvil.

   (*Note:* The bends in this project can be hard to follow. If you become confused, refer to the finished photo to help you figure out the bends.)

2. Seven inches further into the bar, bend the bar back on itself loosely to become the hardy shaft, as shown. Make sure that the shaft is long enough to stick out below the heel of the anvil so that a wooden wedge can be used to wedge it in place. Leave a gap because you will be wedging a piece of wood between the two sides to wedge the fuller in place.

   Bend the bar back 90 degrees, even with the other side of the hardy shaft, to make a "T." I find it easiest to do this part when the shaft is in the hardy.

3. Bend the bar back around, 3 inches further in from the hardy shaft, using your anvil horn. You want the two ends of the bar to run parallel to each other.

4. Clamp the bar in the vise 4 inches back from the end and twist as shown. Repeat on the other end, and you have a matching pair of fullers. You can make fullers of different radiuses by changing the thickness of the bar that you use; just remember to round the edges.

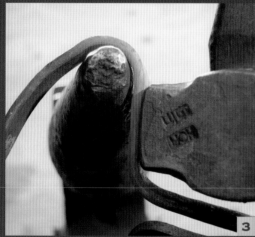

7

PROJECTS FOR THE FARM

100 NAILS PER HOUR

While it may seem simple, making nails is one of the toughest things to get right repeatedly and consistently. Once you can make them in one heat, work toward making the average of 100 nails per hour. To do so, you will need to have many irons in the fire—just don't burn them!

N ow that you have your basic tools made for the forge, you can start making some projects around the farm or yard. Feel free to mix up the different handle styles and twists you will see in the various projects based on your abilities and personal preference.

NAILS, STAPLES, AND RIVETS

Traditionally, nails were made by blacksmiths—it was a good way for them to warm up in the morning and for apprentices to practice their hammer control—or by farmers who were looking to supplement their farm wages. On average, a practiced nail-maker could make 100 nails per hour, heating only once per nail. You will likely need a second heat when you are starting out; I know that I do when I haven't made nails for a while.

HAMMER CONTROL

Making nails is a great way to practice your hammer control. Try to make a nail in one heat with a centered head, and you will very quickly find out how accurate you are with a hammer.

1. Begin by tapering the rod to a point over the near side of the anvil. Set down a shoulder with half-faced hammer blows, as shown. Alternate between 90-degree sides as you hammer to keep the taper square and set down two adjacent shoulders for the header to catch.

2. Once you have a taper, as shown, use your anvil hardy to cut the rod most of the way through 1½ times the diameter of your nail stock above the shoulder. Cut it deep enough so that you can twist the nail off once it is in the header. If you need to reheat the rod to finish the nail head, do so now, while the nail is still partially attached to the rod. Bend it at the cut and place it back in the fire with the nail tip pointed up to prevent burning the nail.

3. Stick the nail into the header and twist it free from the parent stock. Quickly head the nail over the pritchel hole, as shown. Always hit straight down and not too hard; otherwise, the head will fold over, making what is jokingly known as a "corner" nail.

Once it's set down, dress the head by hammering around the edge of the head. Some smiths do a four-sided pattern while others simply hammer around the edge; it is up to personal preference.

NAILS

STARTING STOCK

- Most household nails can be made with ¼-inch rod, preferably square, but you can always square up round rods and/or forge down larger rods.

TOOLS REQUIRED

- Hammer
- Anvil
- Nail header (see project in Chapter 6)
- Hardy hot cut

STAPLES

STARTING STOCK
- ¼ inch round, 3 inches per staple

TOOLS REQUIRED
- Hammer
- Anvil
- Hardy hot cut
- Mandrel or rod to bend around

1. Begin by pointing each end of a 3-inch length of rod into a square taper.

2. Bend around a rod or mandrel of the appropriate diameter, usually ½ to ¾ of an inch.

DID YOU KNOW?

Pre-cut rivets and use a rivet-heading tool if you are making many rivets of the same length.

A rivet is simply a rod that is upset on one end, much like a nail without a point. Rivets can be round, square, or rectangular. Once you assemble pieces with a rivet in place, you peen over the other end of the rivet so that both ends are upset.

1. I like to start upsetting the rod in the vise before moving to the rivet header. You can use a rivet header, as shown in Figs. 2a and 2b, and a spacer to keep the end from folding over, but I find that it often begins to rivet the other end. If the other end does begin to rivet, it can be difficult to get it out of your header. You can then cut off the other end at the needed length, taking into consideration the distance that the rivet will have to span, plus the diameter of the rod for the other rivet head.

2. A rivet header is a plate of ½- to 1-inch steel with multiple holes drilled through it for various sizes of rod (a). Make the holes slightly larger than the rivet stock to allow for expansion when the rivet is heated, generally $1/32$–$1/16$ of an inch. Once you've started the head in the vise, you can then heat the rivet and insert it into the properly fitted hole, with the head formed over the anvil face or clamped in the vise, as shown (b). Once the rivet is headed on one end, cut the rivet to the desired length, which is the width of the assembled pieces it will have to cross plus the same amount you upset on the first head.

THE RIGHT SIZE

As with heading a nail, the length of rod you start with, above the vise, should be 1½–2 times the diameter of the rod—too little, and the head is too small; too much, and it folds over.

RIVETS

STARTING STOCK

- Size and length of starting stock depend on the size of the hole and the width of the metal it needs to span

TOOLS REQUIRED

- Hammer
- Anvil
- Hardy hot cut
- Mandrel or rod to bend around

STRAP HINGES

STARTING STOCK

- ¼-inch x ¾-inch rectangle mild steel, 10 inches long (Adjust the size and thickness of your rectangular steel based on the size of hinge you need.)

TOOLS REQUIRED

- Anvil
- Hammer
- Fuller (optional)
- Cutting plate
- Chisel, hacksaw, bandsaw or angle grinder with abrasive cutting disc
- Vise
- Rivet header

Strap hinges are essential on a farm, and if you have any gates or large doors, this style of hinge is the best to prevent sagging and keep the screws from pulling out. A strap hinge can be as simple as a flat bar with an eye rolled onto the end, as you can find in hardware stores, or as ornate as the beautiful examples seen on gothic churches.

To forge this simple yet decorative style of hinge, I went with a slightly different scroll, choosing an offset double scroll rather than the standard symmetrical scroll. When doing any design such as this, you need to make sure that your end product looks intentionally asymmetrical and not just a poor attempt at symmetry.

1. Taper the tip of the strap over the far edge of the anvil, as shown. The taper should end up 2 inches long.

2. Using a cutting plate to protect your anvil face, with a thin, hot chisel, slit to the start of the taper. In this case, I am planning on an asymmetrical pair of curls, so I start the slit off center at the tip and end on the centerline.

3. Next, neck down behind your slit, as shown, over the far side of the anvil, using full-faced angled hammer blows. Hold the rod at an angle, matching the angle of your hammer face to fuller the metal down. (If you aren't yet consistent with your hammer control, use a set of fullers like I use in other projects in this book.) Fuller the strap down to ½ inch wide by ¼ inch thick and draw out a taper 2 inches long. Chamfer the corners of the taper by holding it so that the opposite corners are in line, perpendicular to the anvil.

4. Roll the tips of the hinge. Start by bending the tip over the edge of the anvil slightly and then hammering it back, as shown. You may need to tweak it to get a smooth

CONTINUED

7

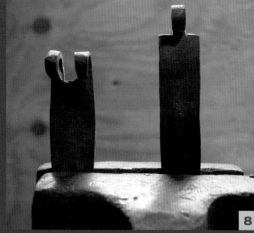

8

9

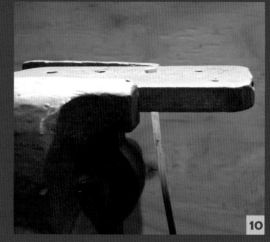

10

ANOTHER METHOD

Another method to forge rivets is to use a ³⁄₈-inch spacer under the hinge to protect the metal for the second rivet head and simply forge the rivet with the hinge assembled. With thin diameter rods such as this, I find it easier to make one head in the vise and finish it in the rivet header rather than making both rivet heads in the assembled project, but you may find it easier to use the spacer underneath the assembled hinge.

HINGE STYLES

There are other styles of hinge that you can make, including wrapping the eye with more length than the diameter of the hinge eye so that you can weld the extra length to the strap in front of the eye for added strength.

scroll, so use light hammer taps to prevent completely collapsing the scroll. Curl the shorter taper more than the longer taper to create an asymmetrical pair of scrolls.

5. Bevel the sides of the strap behind the taper you created in Step 3. Do this over an edge of the anvil to prevent damaging your anvil face. Repeat on the other side of the hinge, keeping the ridge between your bevels in the center of the hinge. Continue beveling until the strap is 1 inch wide.

6. Cut off the strap so that the total hinge length is 9½ inches. Curl the cut end over the far edge of the anvil to begin creating the hinge eye.

7. Continue wrapping the eye around a ¼-inch diameter rod. I usually use a piece of the rod that I'm going to use as the hinge pin for my mandrel to make sure that it fits properly and that the pin can turn freely but snugly. For this one, we're just going to make a butt hinge, as shown.

8. Repeat tapering, beveling, and rolling the eyes on the back side of the hinge. The back side of the hinge should have a 1½-inch taper, beveled to 1 inch wide and with a total length of 5½ inches. Cut both hinge eyes into thirds so that they fit snugly but still rotate easily on the pin.

Using a chisel that matches the width of your cuts—⅜ inch in this case—notch the hinge at the base of the eye on your anvil face. I find it easiest to simply lay it down, eye up, on the anvil face and angle the chisel in under the eye. Once it is notched, clamp the hinge in the vise at the base of the eye and punch the section of eye that you need to remove; it should end up looking like the photo. The decorative strap should have one section in the middle while the back part of the hinge should have one on each side.

9. Make a rivet out of ¼-inch rod by upsetting one end, as shown. If you don't have a pritchel hole on your anvil, use a rivet header clamped in the vise, as I have done. Figure out the distance that the rivet has to span and add on 1½ times the diameter of your rod—⅜ inch in this case—to get the total shaft length; then cut off the rivet.

10. Next, heat up the rivet and quench the formed head and half of the shaft; assemble the hinge and then rivet the second side of the rivet. Use light hammer blows straight down to prevent the rivet from folding over or buckling inside the hinge eyes.

GATE HOOK

STARTING STOCK
- ¼ inch square, 8 inches long

TOOLS REQUIRED
- Hammer
- Anvil

1. Over the far edge of the anvil, draw out a sharp-pointed square taper that is 3 inches long on one end. On the other end, draw out a 3-inch taper that is ³⁄₁₆ inch square at the end.

2. Shown here is the finished taper that will become the hook. The end that will become the eye should be less pointed.

3. Over the far edge of the anvil, bend 3 inches of the hook end 90 degrees. This should put the bend at the start of the taper.

4. Bend the hook around your mandrel, your anvil horn, or a ½-inch-diameter rod. Start the bend ½ inch from the corner that you created in the previous step. Only bend the tip around until it is straight and perpendicular to the shaft of the hook.

5. On the other end of the hook, bend the end of the rod 90 degrees over the far side of the anvil, this time using 3½ inches of material. Bend the eye of the hook completely around the mandrel this time so that the end butts up

CONTINUED

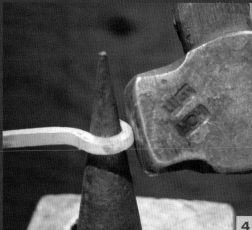

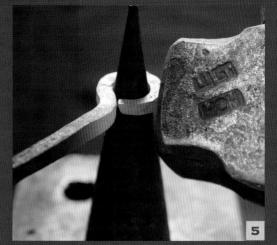

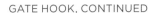

to the offset, as shown. If it is not lined up so that the eye is centered with the shaft, lay the eye over the far edge of the anvil with the offset pointing up and lightly tap it down back into line. Make sure to quench the eye to prevent deforming it as you realign; don't feel bad if you have to flip the hook over and tap it back into line because you went too far—it happens to the best of us.

6. Now, on to the decorative twist. This time, we will be using a reverse twist, but don't worry—it looks more complicated to make than it is. Simply heat half of the shaft, clamp it in the vise at the halfway point, and twist it with a pair of tongs or pliers. Remember which way you twisted it because you will be twisting the other half in the opposite direction, which is why it's called a "reverse twist."

7. Heat the untwisted half of the hook and quench the twisted part; this will keep the twisted metal cool and less malleable, preventing you from untwisting it. Make sure that you heat the metal to the same degree as you did when you made the first twist. Now, clamp the hook in the vise with the heated half up and twist.

    Congratulations on your first reverse twist! Now you can use this technique wherever you'd like to add a decorative twist.

    You can mount the hook using a staple through the ring, with a second staple to hook it to.

DRIVE HOOK

STARTING STOCK
- ³⁄₈ inch round, 8 inches long

TOOLS REQUIRED
- Hammer
- Anvil

1. Begin by tapering out a 3-inch square point.

2. Bend the taper 90 degrees over the far edge of the anvil.

3. On the other end of the rod, set down ¾ of an inch over the far edge of the anvil, holding the rod at an angle and using half-faced hammer blows. Rotate the stock, continuing to fuller a divot until you have circled the rod. If you have a fuller, you can use it to create the divot, but this is good practice in using all of your anvil's surfaces.

4. Using the divot you created in the previous step, upset the corner of the end of the rod back into itself, as shown. Continue to refine the end every 90 degrees until you have a dull point.

5. Bend the end around your horn so that you have 2 inches of hook with a 2-inch diameter curve.

1

2

3

4

5

GARDEN HAND SPADE/ TROWEL

STARTING STOCK
- ¼ inch x 2 inch, 10 inches long

TOOLS REQUIRED
- Hammer
- Anvil
- Cross-peen hammer (optional)
- Chisel
- Drill and ⅛-inch drill bit
- Files or sander

1. Begin by spreading out the trowel end either with a cross-peen, as shown, or by holding your hammer so that you hit with the corner of the face. Adjust the width and clean up any bulges that begin to form as you work to prevent folding over the thinned metal. Always thin from the center to the edges to maximize each heat because the thinner edges will cool the quickest compared to the center of the metal. Continue spreading the metal until the trowel end is ⅛ inch thick and 3¼ inches wide by 4 inches long. You can either forge a taper and create your preferred profile while thinning and spreading the steel or forge it as shown and cut your preferred profile into the flattened portion. Creating the profile while forging is quicker but more prone to hard-to-fix errors that than if you cut the profile out.

2. Once you've forged the blade down to its proper dimensions, forge down the handle width over the near edge of the anvil using full-faced blows. In this photo, you can see that there was a curved piece cut out that I will have to cut square before it becomes a cold shut.

3. Once the handle is 1⅛ inch x ¼ inch in diameter and 5 inches long, begin to set down the end to make a decorative leaf ring with which to hang the trowel. You can use a fuller, or, as I have, you can simply use your hammer and the near edge of the anvil. Once you've drawn it out to a ½-inch width, taper the end sharply. This sharp taper will become the tip of the leaf.

4. Set down ⅝ inch back from the tip over the far edge of the anvil with a half-faced blow.

5. Alternate between the near and far edge of the anvil to draw out the stem that will become the loop. Stop before you reach final dimensions to maintain some strength and mass at the junction of the stem and leaf. Often this area cools quickly, leading to cracking as you forge out the leaf. I like to start forging the leaf when the stem is ¼-inch square at this junction.

6. Before finishing off the stem, begin to forge out the leaf over the horn, as shown. This will help spread the leaf wider rather than longer. Try to keep the ridge of metal in the center by working equally on both sides. You may need to refine the tip across the horn with a ball-peen or rounding hammer to keep it from getting too wide.

    Once you are happy with the shape of your leaf, add some texture with a cross-peen hammered on an angle on both sides of the leaf.

CONTINUED

7. Round up the stem and taper it into the leaf. Next, bend it 45 degrees and bring the stem back around a small rod clamped in the vise or a mandrel, as shown.

8. Cut off the corners of the shovel end so that you have a blank like this, ready for dishing the shovel.

9. Using a V swage and a cross-peen, dish the shovel end lengthwise, as shown.

10. Bend the junction of the handle and shovel down slightly, as shown, to create a slight step for comfort in use.

11. Using a cross-peen and either the step on your anvil or a block of steel on the anvil face in the same configuration, straighten the shovel tip back up.

12. Once you've forged the shovel, it is time to put a handle on it. Cut two matching ⅜-inch hardwood blanks as you did for the screwdriver (see project in Chapter 6). Drill two holes, large enough for a standard brazing rod, 1 inch on either side of center. Rivet the handles to the shovel handle as you did for the screwdriver.

UTILITY KNIFE

STARTING STOCK

- 7 inches of the handle end of a horse hoof rasp (I used the other part of the rasp for the scissors project in Chapter 8)

TOOLS REQUIRED

- Hammer
- Anvil
- Oil for quenching
- File or grinder to sharpen edge

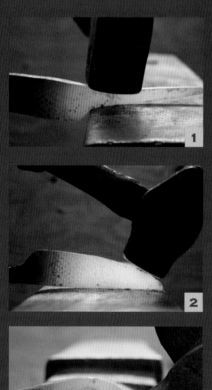

1. Begin by setting down the handle end of the file over the near side of the anvil until it is ½ inch wide. By using the handle end of the file, you save yourself a lot of forging. Make sure to knock back the sharp corners where the file handle begins to create a taper; this prevents cold shuts from forming. Draw out the handle taper until it is 8 inches long, continuing to set down further up the file if you need more length to your taper.

2. Draw out the blade of the file until it is 1⅜ inches wide at the shoulder of the handle. Create a one-sided taper by holding the spine of the knife flat on the anvil and holding your hammer at an angle, as shown.

3. Bend the handle over your horn in the middle and bring the tip completely around so that it is flat against the spine side with a loop at the end. File or grind the blade evenly on both sides until it is approximately ¹⁄₁₆-inch thick. Don't grind it too thin before hardening; otherwise, it may become brittle and crack.

   Anneal the knife by heating it to nonmagnetic and letting it cool in a bucket of ash. After annealing, heat back to nonmagnetic and quench the blade in oil. Temper the blade to straw yellow after repolishing the edges. I find this easiest to do by holding the blade upside down and waiting until the blade shows a straw yellow color and then quenching again.

STARTING STOCK
• ³⁄₈-inch square, 27 inches long

TOOLS REQUIRED
• Hammer
• Anvil
• Drill and ¹⁄₈-inch drill bit
• Files or sander

1. Divide the rod into thirds and cut three quarters of the way through at each mark. Heat the metal to a yellow heat and bend each in half so that you end up with a stack of three bars still connected. Keeping them together like this makes forge welding easier because you don't have to wire or tack-weld them together. Once you've welded together the handle end, you will cut the rake-tine ends apart to work on them.

2. Forge-weld one end of the stack of metal, held on edge as shown. In this photo, I have already split the rake tines and begun working on the leaf decoration. Forge-weld all but 4 inches of the bundle, which will

CONTINUED

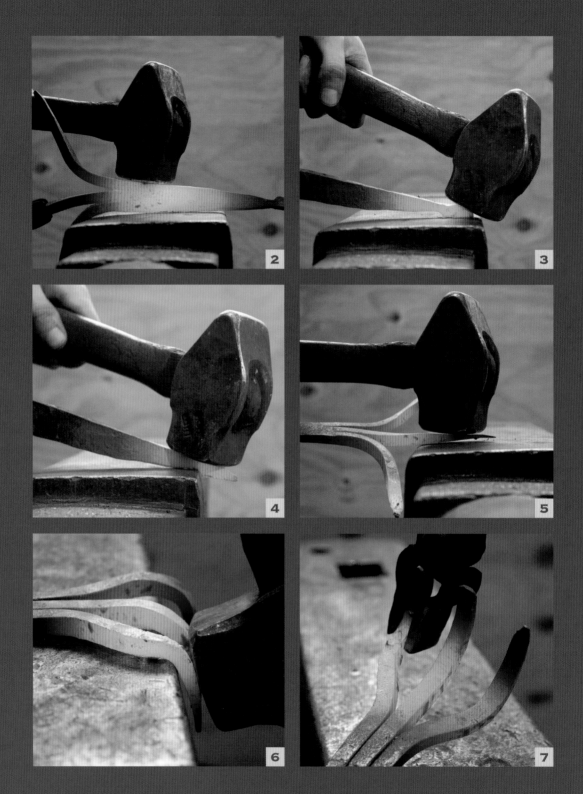

become the rake tines. Continue forging the handle to shape at a welding temperature to prevent breaking your welds.

Over the far edge of the anvil, draw out the handle and flatten it until it is 5 inches long by 1 inch wide and ¼ inch thick. You will need to correct the width to keep it at its starting width of 1 inch. Draw out a sharper taper at the end of the handle so that the end is a ¼ inch square.

3. Taper the tip, rotating 90 degrees to keep it square. (Ignore the fact that I've started necking and isolating the leaf section in this photo.)

4. With half-faced blows over the far edge of the anvil, begin to neck down 1 inch back from the leaf tip. The end dimensions of the leaf transition will be ⅛-inch diameter round, but you will need room to clean up the square corners, and it is always nice to have extra material to prevent it from breaking as you forge the leaf.

   Neck down two adjacent sides by rotating the stock 90 degrees. Finish drawing out the taper, forming the leaf, and bending the ring as you did for the trowel.

5. Bend 4 inches of the outside tines 90 degrees from the middle one, as shown, to give you access to all three tines. Forge the tines into a gradual taper 2 inches long. You will need to trim back the middle tine once you get the outside tines bent to their proper shape. I like to wait until I have everything where it should be and then cut off extra metal to ensure that I don't have a short middle tine, which would then require me to shorten the two outside tines.

6. Using the vise, clamp the handle where you want the bend to occur and then bend the outside tines back parallel with the middle tine. I like to use the vise and a pair of flat-bit tongs to help control where the bending occurs. I've sometimes used the heel of my anvil instead, but I find that I usually end up making more corrections and fiddling with it more.

   Once the tines are bent back parallel to the middle tine, bend all three tines over the edge of the anvil at the same position. In this case, I bent the tines 1 inch forward of where the outside tines were bent back parallel so that there were 1½ inches of outside tine bent over.

7. Now all that is left is to tweak the tines so that they line up. For this, I used a pair of tongs while holding the handle of the rake with another set of tongs. You could use the vise if you have a lot of tweaking to do. After everything is where you want it, cut all three tines off even with a hacksaw. Reheat and taper out any of the ones you cut to match the uncut ones. Sometimes you will need to repeat cutting off the excess and tapering until you are happy with the results.

   After you have the rake forged, make a handle for it as you did for the garden trowel and screwdriver.

HOOF PICK

STARTING STOCK
- ½ of an average full-sized horseshoe

TOOLS REQUIRED
- Hammer
- Anvil
- Softened ball-peen hammer
- Eye punch and thin-bladed chisel

I've been taught to forge the decorative horse head in this project with both the edges of anvils and with hardy fullers. The reality is that both are the same when you look at the shapes that will be moving the metal.

1. Set down and draw out 3 inches of the cut end of the shoe on the near side of the anvil. It should be ⅜ of an inch square at the shoulder and taper to a slightly blunt chisel end. Generally, you can set down at the first nail hole.

2. Using shearing blows over the far side of the anvil, set down the shoulder as shown. Hold the shoe so that the shoulder is facing up, and bend the taper down.

3. Bend the taper back around, over the tip of your anvil.

4. Now that you've made the pick, it's time to forge a decorative head into the toe. Set down a ½ inch of the tip of the shoe's heel on the near side of the anvil with half-faced blows.

5. Set down the neck, 1½ inches back from the nose, over the far edge of the anvil with a half-faced blow until the neck is ½-inch wide. After setting it down, hold the shoe on an angle, as shown, to chamfer the edges and create a rounded profile to the neck.

6. Holding the shoe at a sharp angle, define the jaw and bottom lip; otherwise, you will have a straight nose that won't look quite right.

7. Chamfer the corners of the jaw by holding it at an angle, as you did with the neck.

8. Create a divot to help round out the cheeks. Place the ball-peen (annealed to prevent chipping) where you want the divot and make the divot by striking the ball-peen with the hammer.

CONTINUED

MAKING A NOSE AND EYE PUNCH

To make a nose punch, simply round the tip of a small-diameter punch into a dull ball. Make an eye punch by clamping another punch in the vise and using a nose punch to create a divot in the middle of the punch face. File the edges into either a circle or an oval around the divot.

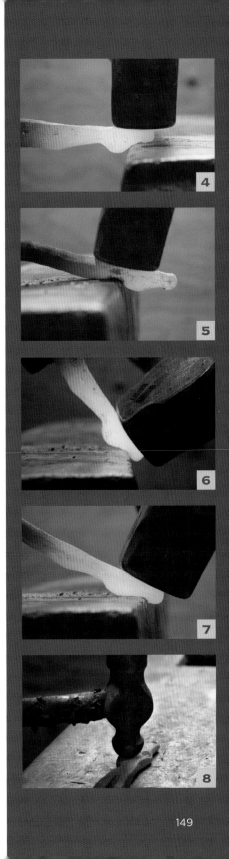

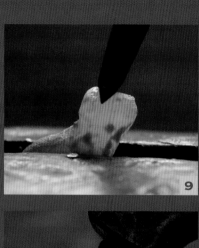

9. Clamp the nose in the vise and cut the mouth with a thin-edged chisel.

10. Cool the nose to protect it from damage and then bend the head around, starting it over the far edge of the anvil and bringing it back around by hammering it, as shown.

11. Over the horn, fuller the mane out wider.

12. Using a fuller perpendicular to the neck, add some texture to the mane to simulate hair.

CHAIN AND HOOK

STARTING STOCK

- Chain loop: 3/8-inch round, 7 inches long
- Hook: 1/2-inch rod, 7 inches long

TOOLS REQUIRED

- Hammer
- Anvil
- Swage
- Punch
- Borax welding flux

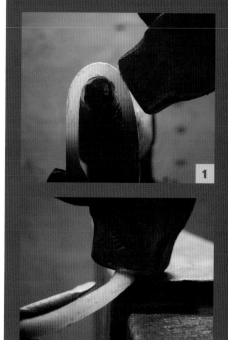

This project shows you how to make a decorative chain and hook, not one meant for heavy towing or lifting. For your safety, please use properly rated chain for any uses other than decoration.

1. Begin by bending the chain link length in half, so that the arms of the U are parallel, 1 inch apart.

2. Over the corner of the anvil face, scarf the ends of the rod into a taper. This scarf will overlap the full-length rod when you forge weld.

3. Bend the ends so that they overlap, as shown. Spread the ends enough so that you can fit another chain link through before forge welding. Generally, chain was made in sets of three links, joined together progressively. Each set of three would consist of two already welded links connected by an unwelded link. Any more than three links becomes unwieldy in a forge fire. To make longer chains, smiths would

CONTINUED

then add a set of three to the longer length with an unwelded link.

4. To make the hook, upset the end of a 7-inch-long piece of ½-inch iron until it is 1 inch in diameter for 1 inch from the end. In this case, I am using a swage to help control the upset, and I make it upset more quickly by forcing the metal in toward the center rather than spreading as much.

5. Flatten the rod to ⅜ of an inch and round the upset over the edges of the anvil (as you will also do in the andiron project in Chapter 8).

6. Taper the tip of the hook end to a dull square point, as shown.

7. This is how the hook should look when it's ready to punch. Punch the hole with an undersized punch and drift the hole to a ⅜-inch diameter. This will maximize the amount of sidewall around the hole.

8. Bend the tip of the hook around so that there is a ⅝-inch gap and so that the tip is just about in line with the beginning of the swelling.

9. Assemble the hook and two chain links, and forge-weld the links. To forge weld, build a deep fire and heat the area where it overlaps to a bright orange. Once it's bright orange, bring the chain out of the fire, wire brush it clean, and cover the weld area with borax flux. Return it to the fire and bring it up to a bright yellow heat. Quickly forge-weld the overlapped scarfs with light hammer blows to prevent shearing the pieces away from each other before they weld.

GATE BOLT LATCH

STARTING STOCK

- Latch plates: three pieces of 11-gauge (⅛-inch thick) sheet metal, 1¾ x 3 inches
- Handle: 12 inches of ⅜-inch rod

TOOLS REQUIRED

- Hammer
- Anvil
- Thin-edged chisel
- Square and ball-nosed punches
- ⅜-inch round drift

1. After cutting or filing the latch plates into elongated ovals, punch two nail holes, 2 inches apart, on center in each. I went with square nail holes to use nails that I forged, but you can use a round punch if you wish to use a screw or manufactured nails. With a 1-inch wide, thin-edged chisel, make two slits, ½ inch apart, to be drifted into a loop.

2. This photo shows the tip of the square-ended nail punch that I used. To make it into a round punch, knock the corners down into an octagon and then round the punch. File it smooth if required.

3. Using a ⅜-inch round drift on one of the latch plates, begin to open up the loop in the pritchel hole (or the hardy hole if the pritchel hole is not large enough).

4. Once you've opened up the loop enough to get the tip of your drift in, begin to drift out the loop, alternating sides to keep the hole uniform and not tapered between sides.

5. The backside of the latch plate will bend around the drift slightly as you open up the loop. Flatten it back down over the edge of the anvil and continue until the ⅜-inch rod slides through the loop smoothly while the backside of the latch plate is flat.

CONTINUED

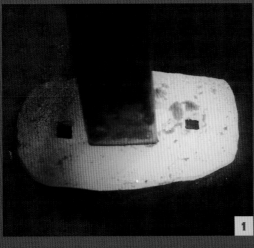

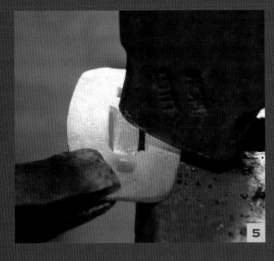

6. A finished latch plate. Make the other two to match.

7. Make the handle by setting down ¾ inch from the tip over the far edge of the anvil, as you did for the drive hook. Continue to rotate the rod, setting it down with half-faced blows until you have enough of a shoulder to catch. Again, you can use a spring fuller, but once you learn how to do it over the edge of the anvil, it will save you much time. Once you have a shoulder, forge the corners back into the bar, as shown, to create a point.

8. After you've upset the tip, begin to taper the neck behind it. Keep the mushroom over the far edge of the anvil to protect it as you taper. Remember to taper it square first and then round it.

9. Bend the handle 90 degrees over the anvil horn, 2 inches from the end of the rod. With a small ball-nosed punch, create a divot 6 inches back from unfinished end of the bolt. Force the metal on one side down with the punch by angling the punch toward the outside of the rod. This will create a lip to catch the latch on. Make sure that it is 90 degrees to the handle so that when the latch is closed with the bolt handle down, the lip is engaged on the latch plate but can slide through when the bolt is turned. You can see this in the photo of the finished project.

8.

PROJECTS
FOR
THE HOME

DOOR HANDLE/ DRAWER PULL

STARTING STOCK
- $3/4$ x $1/4$-inch bar stock, 12 inches long

TOOLS REQUIRED
- Hammer
- Anvil
- Chisel
- Drill or punch

Door handles and drawer pulls can be made in many different sizes, with different chisel designs and designs for the finials. The size I've made is a nice size for closet or pantry doors.

1. Begin by tapering both ends of the bar to a 1-inch taper on the far side of the anvil to protect the anvil face from the edge of your hammer.

2. Beginning 2 inches from the tip, neck down and define where the finial will be. To do this, hold the bar on edge at an angle and match the angle with your hammer face, as shown. Flip the bar to keep the taper symmetrical. Make sure that your hammer blow hits in line with the edge of the anvil. Again, you can use a set of spring fullers, but you should work up to using just your hammer and anvil. Make sure to leave enough width to maintain strength and prevent cracking at the neck as you forge out the leaf shape. Final dimensions are $3/8$-inch wide and $1/4$-inch thick, so taper down to $1/2$-inch wide and finish it only once the leaf is forged to shape.

3. Over the edges of the anvil, to protect your anvil face, taper the handle edges so that the handle is 1 inch wide. Make sure to keep the edges symmetrical. Do the same for the leaf finials so that they are as wide as you can get them. You can use the horn or a cross-peen to help fuller the metal to the side.

4. Using a chisel, cut two 4-inch-long parallel lines, centered in the length of the pull, into the handle.

CONTINUED

1

2

3

4

5. After widening the leaf, you may need to point the tip on the horn. Once you are happy with the leaf's shape, clean up and forge the tapers to their final dimensions as outlined in Step 2.

6. Use the edge of the anvil to bend the tip of the leaf as needed to suit your taste.

7. Bend the neck just past the leaf, 90 degrees over the edge of the anvil, so that the leaf points up when the chisel lines are facing up. Repeat on the other side.

8. Bend both sides of the handle over the horn, as shown, where your chisel lines begin.

DIY GIFTS

The great thing about blacksmithing is that you can create custom ironwork for your house and as gifts for friends and family, even if you never sell anything to another person. This chapter features some of my favorite projects and popular items for around the home.

DINNER TRIANGLE

STARTING STOCK
- Triangle: 1/2-inch diameter rod, 36 inches long
- Beater: 1/2-inch diameter rod, 14 inches long

TOOLS REQUIRED
- Hammer
- Anvil
- Cross-peen hammer
- Blunted chisel (to use as a fuller)
- Punch
- Scrolling tongs (optional)

1. To make the triangle, begin by flattening the end of the rod into a fishtail scroll, as described in Chapter 4. Make sure to keep the widening even.

2. Using your cross-peen, widen the fishtail even more.

3. Over the far edge of the anvil, taper just before the scroll starts to add more elegance to the scroll.

4. Over the near side of the anvil, clean up and even out the scroll.

5. Using a dull chisel, fuller three parallel lines into the scroll to add more definition. One line should be centered, with the other two being parallel to the edges.

6. Begin the scroll over the far edge of the anvil and

CONTINUED

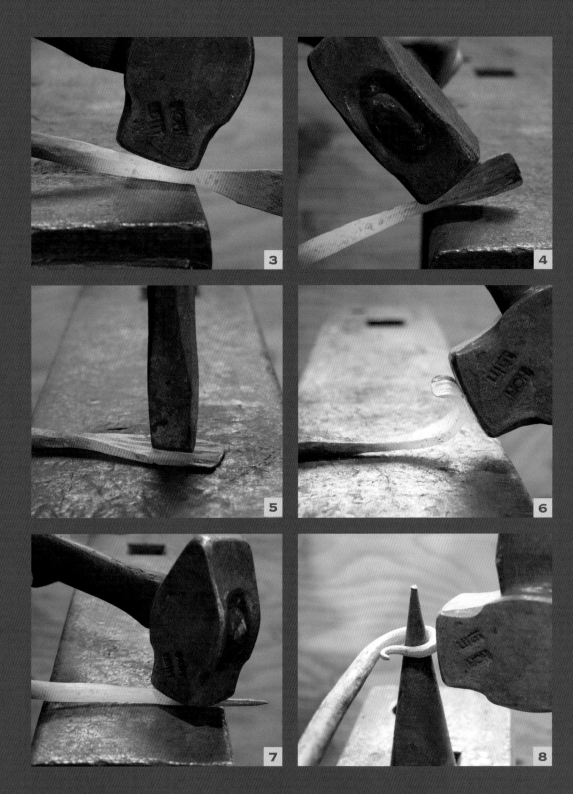

continue rolling the scroll back, as shown, until you scroll into the narrowed taper.

7. Taper and round the other end of the triangle so that you have a 5-inch taper.

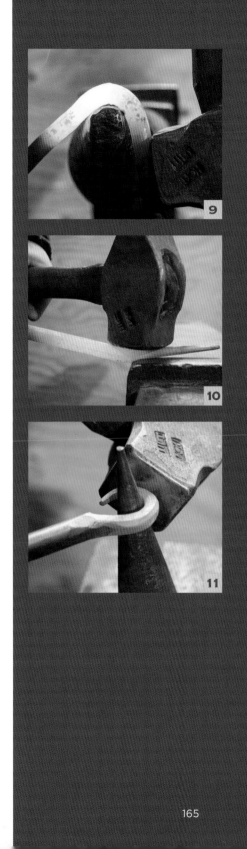

8. Curl the tip either over the edge of the anvil or with a pair of scrolling tongs; bend the end back around, as shown. You can either use a mandrel, as in the photo, or a pair of scrolling tongs if you aren't advanced enough to use your anvil horn and close it over the anvil face.

9. Mark the triangle into three equal sections with a center punch and bend the rod over the horn, at the two angle marks. It should make an equilateral triangle.

10. To make the striker for the triangle, set down two adjacent sides, 1½ inches from the tip, over the near edge of the anvil. Taper and round the tip.

11. Next, bend the taper 45 degrees away from the shoulder you created and bring the tip back around to the shoulder, as shown.

COAT HOOK

STARTING STOCK

- ½-inch-square rod, 10 inches long

TOOLS REQUIRED

- Hammer
- Anvil
- Softened ball-peen hammer
- Sharp and dull chisels (to use as fullers)
- Brass brush

1. Upset one end of the rod over the anvil edge, as shown. Use light blows to rivet the end over until it is ¾ of an inch square.

2. Taper the other end in a steep 1½-inch taper and set down two adjacent sides over the far side of the anvil at the end of the taper, as shown. Don't take the shoulder down to its finished dimension (⅜–inch round); it can cause cracking because the shoulder will cool faster than the leaf. For now, don't go further than half of the metal's thickness.

3. This is how the leaf should look with both shoulders set down.

4. Flatten the leaf on the diamond so that the corners will push into the leaf and, in effect, fuller the leaf wider.

5. Using the horn as a fuller, spread the leaf wider. Alternatively, you could use a cross-peen hammer.

6. Finish tapering the stem for the leaf. Knock the corners off at the start of the taper and proceed to round the stem.

CONTINUED

7. This is the finished stem. Notice the progression from square to octagon to round.

8. Strike the softened ball-peen (make sure that it is softened to prevent flying chips) with your hammer to create two divots, 1 inch apart, starting approximately 1 inch from the start of the taper. These will become the countersinks for mounting the hook to the wall. Simply drill out the center to match your preferred screw size.

9. Bend the upset end around the horn of your anvil, 1½ inches from the end.

10. Over the corner of your anvil face, to protect the divot and upset end, correct any widening of the bend that will occur.

11. With a chisel, cut a line down the center of the leaf. Make the veins rise up by using a small fuller to set down the metal between the veins. Feel free to experiment with different fullers and textures until you find one you like.

12. Using a ball-peen, dish the hot leaf from behind on the end of a small block of wood. The metal will burn down into the stump slightly.

    To add a brass finish to the leaf, brush the leaf with the brass brush as it cools down from cherry-red heat.

ALTERNATE COAT HOOK FINIAL

1. Begin by setting down ¾ inch over the far side of the anvil with half-faced hammer blows. Continue doing the same around the bar.

2. Once you have a shoulder to catch on the anvil, begin forging the corners back into themselves. Rotate the bar 90 degrees to keep things square.

3. Continue to refine the taper into a four-sided pyramid and taper behind the pyramid. In this photo, you can see a crack forming, which spells disaster for this coat hook. Apparently, my scrap was made of other scrap that had been weakly forge-welded together.

BBQ FORK

STARTING STOCK
- Fork: ³/₈-inch-diameter rod, 20 inches long
- Handle: ¼-inch-diameter round, 32 inches long, folded into quarters

TOOLS REQUIRED
- Hammer
- Anvil
- Hardy
- Wire brush
- Borax flux

This project will test your forge-welding proficiency. You can substitute the handle style with one from any other project in the book if you aren't able to drop-tong weld yet, or you can "cheat" with a welder to tack the pieces together. I would encourage you to practice drop-tong welding, though, because you never know when your welder might die the night before a project is due for a client.

1. Begin by folding a short length, approximately ³/₈–½ inch, over the far edge of the anvil to 90 degrees. Hammer this corner back into the parent stock, as shown, to upset and create the scarf for welding.

2. Forge the tip into a sharp point so that the scarf blends in nicely when welded and you don't end up with a large shearing plane that will be a weak point.

3. The finished scarf, upset at the welding point, and scarfed tip.

4. Bend the rod around the horn or a mandrel, as shown, so that each prong is 3 inches long and the scarf is lined up with the shaft of the fork.

5. Bring the weld and the area around the weld up to near-welding heat, wire-brush it clean, and apply borax flux. Return it to the fire and bring it up to a bright-yellow welding heat to forge-weld it. Remember to start with lighter blows to get the weld to stick and then increase your hammer blows as you dress the weld. Feel free to reheat the weld while dressing it if it becomes too cool.

CONTINUED

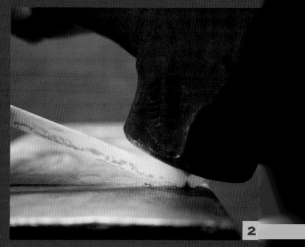

FOOD-SAFE FINISHES

There are many different options to rust-proof your BBQ fork that are food safe; these include beeswax, paraffin wax, vegetable oil, and linseed oil without any hardeners. To use, heat the iron up until the oil smokes to create an attractive blackened finish.

6. Once you are happy with your weld, cut apart the prongs in the middle of the loop.

7. Bend the tines apart and taper them over the near side of the anvil. You can leave the prongs square or proceed to rounding the tapers.

8. After tapering, bend the tines apart symmetrically and then bend them back to parallel around your anvil horn, as shown.

9. Taking your length of ¼-inch rod folded into quarters, forge weld 2-3 inches of both ends together.

10. Upset one end slightly and forge a scarf into the end, over the edge of the anvil. Remember to point the tip so it blends in nicely.

11. This is the finished handle, ready for welding, after tapering the other end into a round taper at welding heat. Bend the taper into a small eye loop to hand the fork.

12. Drop-tong welding is the hardest thing you will do as a blacksmith, in my opinion. It takes a lot of practice to be smooth and efficient enough to get a nice weld. First, heat both ends to welding heat after fluxing. You can test when it is close to welding temperature by touching the ends together in the fire and seeing if they stick together. Holding the tongs with the handle in your hammer hand and the fork shaft in your other hand, quickly move to straddle them on the anvil face. You need to have the longer piece on top so that it will pin the piece in your tong hand against the anvil.

CONTINUED

BBQ FORK, CONTINUED

13. Pin the piece in your hammer hand to the anvil with the second piece and drop your tongs to quickly grab your hammer to begin the forge weld. Once the weld is set, reheat it to welding temperature to finish the weld.

14. Clamping one end of the handle in the vise, twist the handle an even number of turns with a pair of tongs. Make sure to straighten any bends on the anvil before proceeding to untwisting to create the basket twist.

15. Untwist the twist you just made by half the number of turns you did to twist it up.

SHEARS

STARTING STOCK

- Farrier hoof rasp
- Rivet: ¼-inch–diameter rod (length should be twice the rasp's thickness plus ¾ inch)

TOOLS REQUIRED

- Hammer
- Anvil
- Slot punch
- Hardy or chisel

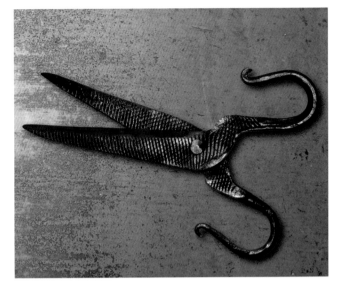

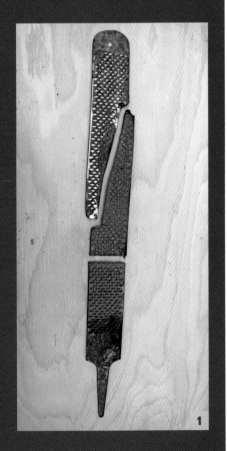

1. This is a farrier's rasp cut into parts for this project as well as for the utility-knife project in Chapter 7. If you plan ahead, you can maximize your stock and minimize waste. The pieces for the shears are 7 inches long, including the handle end, which is a minimum of 1½ inches. All of the following steps are identical for both sides of the shears.

2. With a ¾-inch slot punch, slot into the middle of the end, as shown. This will save you time forging down the stock. I use a slot punch, not a slitting chisel, in this instance to reduce the amount of rag left in the hole that would end up becoming a cold shut later if not filed off. After you slot-punch the hole, use a chisel and a cutting plate or your hardy to cut from the end of your slot punch to where the corner of the notch is.

3. Straighten out the loop of steel you just created. You will need to forge the corners so that they are flush with the rest of the steel. Do this over the near edge of the anvil to protect the jog in the steel, as shown. Taper and round the steel until it is 5-5½ inches in length. Make sure that both sides match for aesthetics.

4. Bend the tip of the taper into a tight scroll, then bend it into a wider loop halfway along the taper. Once you are happy with the forging, and both shear blanks are close to identical, file or grind them until they are as close to perfectly identical as you can get. Grind them sharp from the outside, as you see on any pair of scissors, and make sure that the inside edges are smooth and flat so that they match up nicely.

Drill a ¼-inch hole in the center over the widest point before the handles start and then rivet them together with a ¼-inch rivet. You can see the assembly and rivet placement in the photo of the finished piece. Now tweak the handles until they feel comfortable in your hands, and tweak the blades so that they open and close easily and stay in contact with each other to properly cut. Just like your tongs, you can heat the rivet and open and close the shears a few times to loosen the rivet slightly. Proceed to heat-treating the scissors and tempering to a straw yellow to brown at the cutting edge, depending on their use.

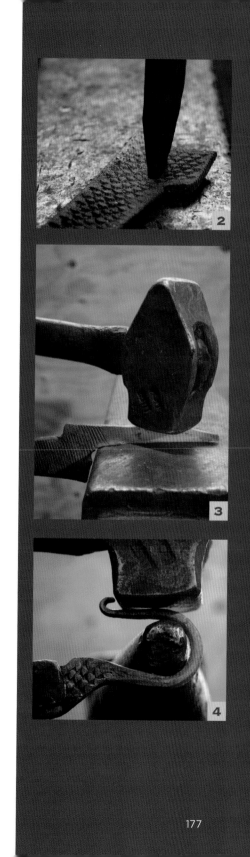

FIREPLACE SET

STARTING STOCK

- ³/₄ x ¹/₄-inch medium-carbon steel, 30 inches long, for each of the three implements: shovel, broom, and poker. I used medium-carbon steel because it will be thinner for appearances and, most importantly, weight
- 18-gauge metal for shovel (what type and dimension)

TOOLS REQUIRED

- Hammer
- Anvil
- ¹/₂-inch slitting chisel
- ¹/₂-inch round drift
- Hacksaw or thin-edged hot cut
- V swage and cupping tool

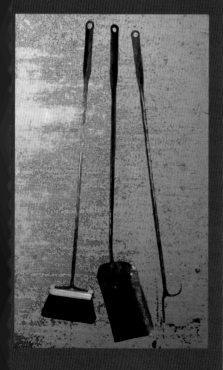

HANDLE AND SHAFT

1. The broom, poker, and shovel begin with the same handle and shaft. Over the near edge of the anvil, taper 6 inches of one side of one end, as shown.

2. Repeat tapering over the far edge of the anvil, as shown. Taper both sides down to ⅛ of an inch on the edge and 1 inch wide.

3. Forge the corners of the end into the bar and round up the end. With a round punch, punch and drift out one end of the rod. With the punch in the bar, round the ends up if required. Drift the hole to a ½-inch diameter.

   Over the near side of the anvil, forge a taper into the end where you spread the handle. Proceed to draw the handle down to ½ x ⅜ inch at the start of the shaft, where it meets the handle taper. Continue drawing the shaft out in a gradual taper to a ⅜-inch square for a length of 23 inches for the poker and broom. Because of the length of the shovel scoop, the shovel shaft needs to be tapered for 19 inches. There will be enough metal left over at the ends to make the different heads.

FIREPLACE BROOM

4. Over the near edge of the anvil, adjust the taper on the broom end of the handle, as shown.

5. Cut off the end of the bar so that there are 1½ inches of stock past the taper. Slit the bar in half up to the end of the taper. Split them apart and forge out the tip until the edge is ⅛ inch thick; then, round up the end with a file. Center-punch and drill a hole in both arms to screw the handle to a broom. I used a small brush with soft bristles that I cut down to size and sanded (see photo at left).

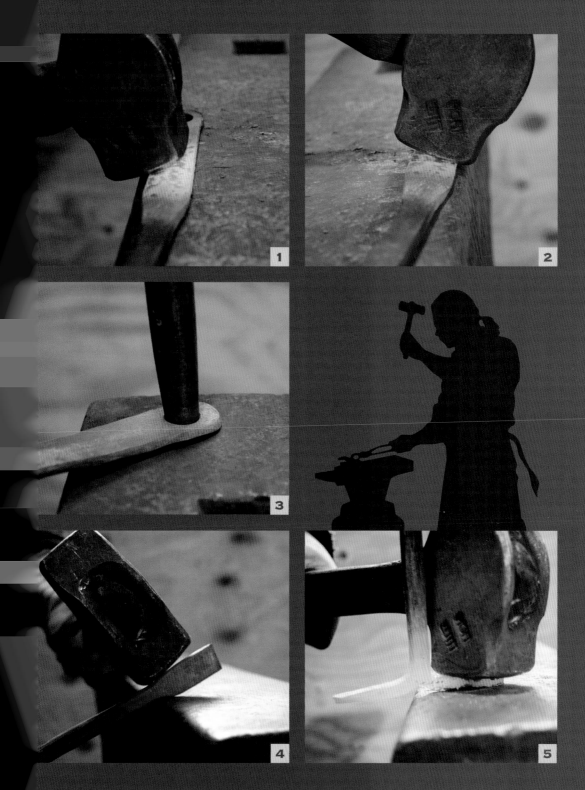

FIREPLACE POKER

6. Create a taper where it widens at the end, as you did for the broom, and cut off the end so that there are 2½ inches of untouched bar at the end. Split it down the middle and forge both legs into a 3-inch round taper.

7. Straighten one tine and bend the other backward over the anvil horn, as shown.

FIREPLACE SHOVEL

8. For the shovel, I welded together a couple of pieces of scrap to obtain the proper length, but you can draw out the bar as you did for the other fireplace tools. To make the shovel rivet holes, begin by slitting two holes, 2 inches apart, with a slitting chisel, as shown.

9. Drift open the holes with a ½-inch-diameter drift. Proceed to make an 8½-inch-long and 6-inch-wide shovel blank out of 18-gauge sheet metal, as in the photo. Using a V swage and a rounding hammer as you did for the coal scoop/scuttle in Chapter 6, slightly dish the edge 1 inch in from the outside.

ANDIRON

STARTING STOCK
- $^3/_8$ x 2-inch bar, 18 inches long
- $^5/_8$-inch-square bar, 25 inches long

TOOLS REQUIRED
- Hammer
- Anvil
- Spring fuller
- $^3/_8$-inch drill bit and drill or punch
- Bandsaw, hacksaw, or hot-cut chisel

This is a traditional-looking fireplace accessory used to keep logs from rolling out of the fireplace while allowing oxygen under the fire.

1. Taper one end of the $^3/_8$ x 2–inch bar stock until it is 6 inches long and 1 inch wide on edge. Flatten the bar to a blunt taper and allow it to widen slightly into a fishtail.

2. Scroll the end into a fishtail scroll (see Chapter 4).

3. On the other end of the bar stock, split 8 inches in half and draw both legs out until they are identical and ¾ inch wide.

CONTINUED

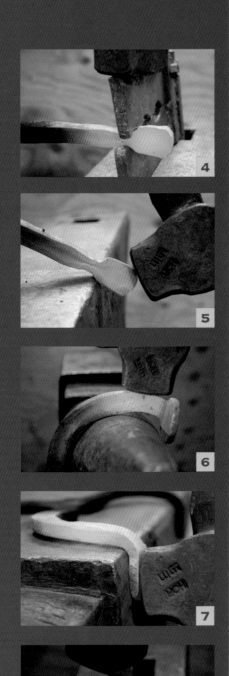

4. Once the legs are symmetrical, use your spring fuller to narrow 1½ inches back from the tip down to ½-inch thickness. Forge both legs until they are a gradual taper to blend in the divot.

5. Using the divot you created with the spring fuller to catch the edge of the anvil, proceed to forge the corners into the center of the mass and round it. Alternate between the near and far edge of the anvil until it is round and then flatten the circle down to ¼ inch.

6. Bend both legs sideways and bring them around the horn, as shown.

7. Bend both feet 90 degrees over the far edge of the anvil just before the foot, as shown.

8. An inch back from the tip, fuller the ⅝-inch square down to ⅜-inch round with a spring fuller.

9. Over the near edge of the anvil, draw and round the tenon to a ⅜-inch diameter.

10. Rather than making a monkey tool for this project, drill or punch your hole in the upright to match the tenon on the leg. With a center punch, mark one side of the bar and one spot on the upright. By matching these marks up at every heat, you can get a better fit when squaring up the shoulder of your tenon. Hammer down on the square bar with the tenon in the pritchel. Continue to upset and flare the shoulder slightly.

11. Using your hardy, cut off the tenon so that it is ⅞ inch long to span the upright and give you enough to form a rivet head.

12. Assemble the piece in the vise, as shown, and rivet the tenon over until tight. Measure the height of the upright from the feet to the tenon and bend that distance back from the end of the square bar 90 degrees to create the back leg.

13. The finished tenon rivet.

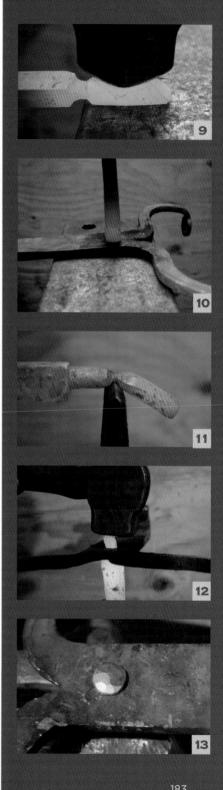

CAMPFIRE TRIPOD AND SPIT SET

STARTING STOCK

- Three ½-inch square bars, 50 inches long

TOOLS REQUIRED

- Hammer
- Anvil
- Narrow fuller
- Borax flux

1. Taper one end of each rod into a 2-inch square point by rotating it 90 degrees between hammer blows.

2. The finished taper.

3. On the other end of each rod, taper one side of the bar and let it widen evenly into a fishtail (see Chapter 4). If you need to widen it more, feel free to use a cross-peen, but remember that it will spread when you fuller in lines later.

4. With a narrow fuller, create three lines up the taper that divide the fishtail into quarters.

5. Begin the tip of the scroll over the far edge of the anvil as shown. Be careful not to pinch the metal fully against the anvil because that will damage your fuller marks.

6. Continue scrolling by rolling it back on itself, as shown.

CONTINUED

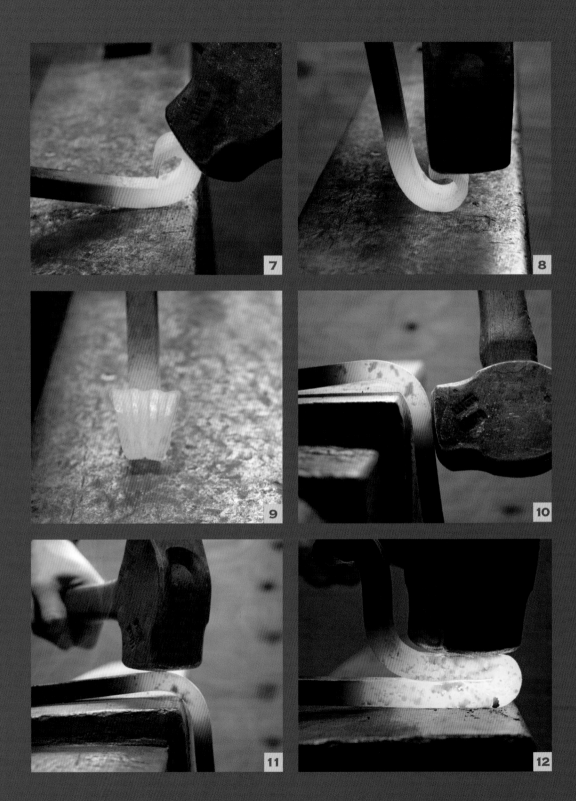

7. Tighten the scroll as needed by rolling the scroll.

8. If you need to tighten the inside of the scroll, hold the bar straight up, as shown.

9. The finished fishtail scroll.

10. The measurements in this step will be different for one of the three pieces, but you should bend all of the pieces with the scroll facing down. For the one that will become the ring, make a 45-degree bend 9 inches back from the end of the bar after scrolling. Now bend the bar back around to make a 2-inch-diameter eye.

Bend the other two pieces back on themselves at 5 inches from the end of the bar.

11. Correct any bowing that occurs as you make your bend.

12. Flux and weld the area where you folded the bar back on itself on two of the pieces.

13. Continue to draw and taper the tips at a welding heat. You want a 2½- to 3-inch matching taper on both pieces. Chamfer the edges and round up the taper so that they can spin easily when assembled into a tripod.

14. Bend the unwelded part into a hook on the two welded pieces, as shown. You want the hook to be 1 inch wide and 1½ inches deep.

15. Chisel a 5-inch line into all four sides of each piece, in the center of the length.

16. Proceed to twist all of the pieces where you chiseled.

CAMPFIRE TRIPOD S-HOOK

STARTING STOCK

- ¼-inch square bar, between 6 inches and 2 feet long, depending on desired distance from fire

TOOLS REQUIRED

- Hammer
- Anvil
- Vise

1. To cook on a fire, make an adjustable trammel or a set of S-hooks of various lengths. I've chosen to make a ¼-inch–square S-hook here with a penny scroll.

2. Begin by setting the ¼-inch square rod ½ inch back from the end, down over the far edge of the anvil, to create a shoulder. Once you have a shoulder to catch, as shown, begin to forge the end back into itself.

3. Continue forging the metal back until it is rounded and ¼-inch wide. Set the neck down to ⅛ x ¼ inch. Bend the penny 90 degrees.

4. Forge the penny down into the neck to create a sharp transition at the neck.

5. Curl the scroll around. Tighten up the scroll as needed by hammering down into the anvil.

6. Bend the hook around a horn or mandrel. As you can see, a rod in the vise can be used instead of an anvil horn. The hook should be 1½ inches deep and 1 inch wide.

7. Add a twist to the middle, and you're finished! Just make sure that the hooks point in opposite directions; otherwise, it is a C-hook, not an S-hook, and it won't hang properly.

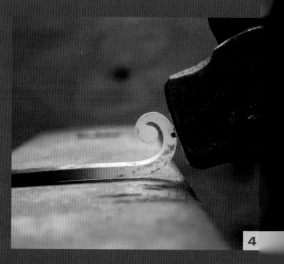

SIMPLE CANDLE HOLDER

STARTING STOCK

- Base: 1/8 x 1-inch bar, 12 inches long
- Dish: 3-inch square piece of 18-gauge metal
- Cup: 1-inch-diameter schedule 40 pipe

TOOLS REQUIRED

- Hammer
- Anvil
- Chisel
- 1/4-inch drill bit
- Pipe flaring tool (the cut-off end of a tire iron works perfectly, but anything with a similar shape will do)

1. Begin by splitting 5 inches of one end of the bar. Remember to use a cutting plate to prevent damage to your anvil.

2. Gradually taper both tips of the ends you split over the near side of the anvil.

3. Taper the other end 7 inches, on edge, into a point.

4. With a cold chisel, incise lines along the two split tapers and the opposite side of the individual taper.

5. Bend the bar 90 degrees, 1 inch before the split ends, over the far edge of the anvil.

6. Curl both split tapers in opposite directions on the horn, as shown. This is curling the hard way, and you will need to flatten the metal back down because it will want to twist.

7. Using your cross-peen over the far edge of the anvil, bend the bar 90 degrees, 1 inch back from your first bend, which should be where the end splits.

8. Over the heel of your anvil (or vise jaw, if it is the right size), bend it another 90 degrees, as shown, 2 inches from your last bend.

CONTINUED

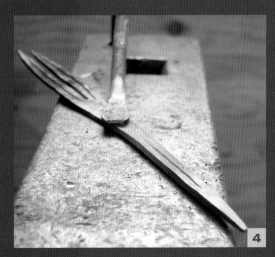

9. Bend the tail back 90 degrees, 1 inch from your last bend, as shown.

10. Curl the tail around on the horn to make a decorative handle.

11. To prevent hooking yourself on the point, curl the tip back around on a smaller mandrel or with a pair of scrolling tongs.

12. Curl the corners of your 3-inch-square piece of sheet metal by starting them over the far edge of the anvil and dressing the curl as shown.

13. Dish the candle cup slightly concave, raising the corners slightly.

14. Holding the pipe at a steep angle, hammer the end in on itself, rotating the pipe between each hit. Neck it down to a ¼-inch diameter for the rivet. *Important*: Plug the end of the pipe with a rag to prevent getting burned by heat rising up the pipe or steam if you quench it.

15. A flaring tool, made from a cut-off tire iron.

16. Cut off the pipe 1 inch long and hammer it down onto the flaring tool.

17. Drill a ¼-inch hole in the center of the platform you created with the bends and then assemble and rivet the cup and dish to the base using a rivet with one head made. With a bull-nosed tool, longer than the depth of the candle cup to support the rivet head inside the cup, rivet to the other side on the bottom of the base.

15

16

17

LEAF-SHAPED INCENSE HOLDER

STARTING STOCK
- 1¼-inch angle iron, 10 inches long

TOOLS REQUIRED
- Hammer
- Anvil
- Hacksaw, bandsaw, or angle-grinder with a cut-off disc
- V swage

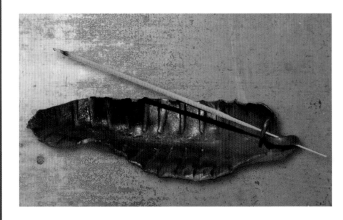

1. Begin by cutting out the angle iron to the approximate shape shown in the photo. The main part of the angle iron should be 8 inches long, and the tip should be cut with a slight angle to both sides. Notch down so that there is about a ½ inch of angle iron left on each side.

2. Treat both sides as a separate bar and taper both sides of the notch into a square point. Round the square section of the taper and chamfer any edges where it is still flat.

3. Forge both corners of the notch round. Remember that angle iron is essentially two flat bars when it comes to forging.

4. Forge the tip into more of a point, as shown.

5. Flatten the angle iron and forge the corner flat.

6. Using a fuller with a diameter of your choice, incise some lines, as shown, or leave it plain.

CONTINUED

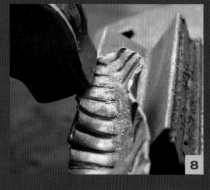

7. Using a V-swage, dish the flattened piece into a curled leaf shape.

8. Roll the edges over on the wider parts you have dished out.

9. With a pair of scrolling pliers, bend the taper up and around into the dish.

10. Create a coiled loop to hold the incense.

SHELF BRACKET WITH A SQUARE CORNER

STARTING STOCK
- ¼ x 1¼-inch bar, 16 inches long
- ½-inch square bar, 10 inches long

TOOLS REQUIRED
- Hammer
- Anvil
- ¼-inch drill bit and countersink
- Swage block (optional)

1. Begin by upsetting both ends into a noticeable flare. Over your horn, flatten the top of both ends down so that the flare is down and it is flat on top. This is where the wall and shelf will sit.

2. Over the far edge of the anvil, bend the bar 90 degrees at the halfway point.

3. Bend the ½-inch square bar 90 degrees at the center. Begin to upset the corner so that you can forge a square corner into the bend by hammering down on the leg. Repeat on the other leg to evenly upset both sides into the corner.

4. On the far edge of the anvil, upset the corner into the leg more by hammering it, as shown. Make sure to leave a small gap between the bar and the anvil to prevent drawing it out. Once you have enough mass, flatten the sides back to ½ inch and forge the corner square on the far edge of the anvil. Unless you have a sharp 90-degree corner on your anvil or a swage block, the inside corner will have a curve matching the radius of the anvil edge.

5. Bend both ends around on the horn, as shown, and flatten the bar into a taper. Now line up and drill a ¼-inch hole in the middle of each taper on the brace and the shelf bracket. Countersink the side of the bracket that will be in contact with the wall and shelf to create a hidden rivet.

   To make a hidden rivet on the back of the shelf bracket, use a normal rivet and rivet the end into the countersunk hole. After filling the hole, file or grind it smooth and flush with the back of the bracket.

DOOR LATCH

STARTING STOCK

- Handle: ½-inch square bar, 12 inches long
- Thumb latch: ¼ x 1-inch bar, 4 inches long
- Latch staple: ¼ x 1-inch bar, 4 inches long
- Latch bar: ¼ x 1-inch bar, 8 inches long
- Latch catch: ¼ x 1-inch bar, 5 inches long

TOOLS REQUIRED

- Hammer
- Anvil
- Cross-peen hammer
- Punch and drift for a ³⁄₈-inch hole

1. Begin by tapering one end of the ½–inch bar into a sharply angled square point to make a spade.

2. Over the far edge of the anvil and 2 inches back from the tip, set down adjacent 90-degree sides to ³⁄₈ inch with half-faced blows, as shown.

3. An inch behind your first shoulder, set down another shoulder on the near edge of the anvil to isolate the material between the latch and the spade. Make sure that it is set down to only ³⁄₈ inch to start and on one side. We will finish forging the stem that we're creating once the spade is finished.

4. An inch behind the last shoulder, set down another shoulder with half-faced hammer blows. Again, this is only one side, and it will isolate a raised boss to be slot-punched for the thumb latch to be riveted into. Continue to flatten the square bar to ³⁄₈ inch thick x 1 inch wide for 8 inches. You may need to use a fuller to widen the bar.

   At the end of the flattened portion, create a shoulder as you did in Step 1 and cut off all but 1½ inches. Repeat the first two steps on the square bar to make a spade on the other end of the handle but without the boss.

5. Using a ¾-inch slot punch, and alternating sides to get a less tapered slot, punch and drift the hole for a ³⁄₈-inch wide x ⅞-inch-long thumb latch with parallel sides. This photo was taken after I completed forging the spade and stem and needed to square the hole back up.

6. This is the latch, ready to have the spade forged and the stem cleaned up. As you can see, the hole ended up slightly off center, but that's not a problem. If you start with an undersized punch, you can forge one side more than the other, after cooling the longer side, to equalize the hole.

CONTINUED

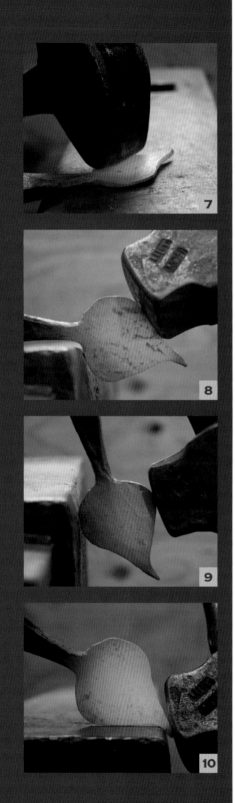

7. Using your cross-peen or anvil horn, fuller the spade wider so that it will have better stability when mounted on the door.

8. Refine the shape over the far edge of the anvil, as shown, which will upset more metal into the bulge of the spade.

9. Continue to refine the roundness until it is aesthetically pleasing.

10. Draw out the tip of the spade and curl it slightly over the far edge of the anvil.

11. Refine and center the taper on edge. Remember that it is only slightly thinner on the flat than on edge to maintain its strength when being pulled on.

12. Continue to refine the taper and forge the stem down to its final width.

13. Chamfer the edges of the handle section for comfort by holding the metal on its corner and forging the corners back into themselves, as shown. You can either leave the handle blank or fuller a decoration into the handle portion before the tapers start, as I did.

14. On the end with the latch boss, bend the handle 90 degrees just before the boss. Bend the other end just before the spade so that both spades are pointing in the same direction as the fuller lines in the handle.

15. Bend both sides of the handle at the start of the taper over the horn with your cross-peen hammer, as shown.

16. Make a staple, which will keep the latch bar from flipping up too far. Draw out some scrap until it is ½ inch wide and

CONTINUED

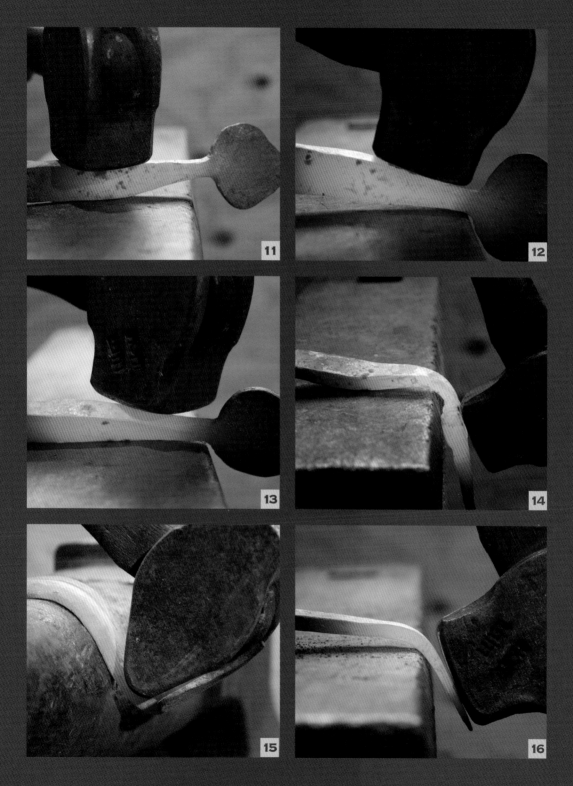

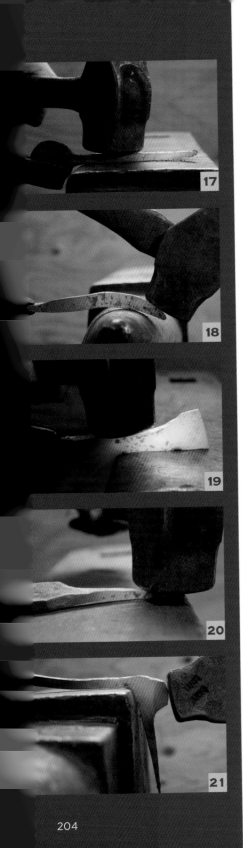

¼ inch thick. Taper one end into a 2-inch-long square point and, leaving 2½ inches untouched, taper the other end into another 2-inch-long square point. Cut off any excess and bend both tapers 90 degrees in the vise to create a square staple.

17. Leaving 1½ inches, forge the bar for the thumb latch on edge until it is ¼ inch wide.

18. Draw the end out to ½ inch wide and 5 inches long. Taper the tip into a dull point and curve it down, as shown.

19. Forge one end of the stock for the latch catch into a 3-inch-square taper that is ½ inches long x ¼ inches wide at the shoulder. To create the shoulder, use half-faced blows on the near edge of the anvil. Repeat the same on the other end, leaving 1 inch of the stock untouched between the shoulders; forge the leg into a uniform ½ x ¼-inch cross-section. Cut it off at 2½ inches.

20. Remember to use the far edge of your anvil when tapering to a sharp point.

21. Bend the untouched mass over the far edge of the anvil on the untapered side, as shown. Clamp the metal in the vise with the bed facing up and a gap between the inside bend and the vise jaws; finish forging the corner into itself so that the bend is at the untouched stock and not on the forged-down section.

22. Forge a slight bevel into the catch, as shown, so that the latch bar can slide up and over properly.

23. Using a small mandrel, bend the untapered leg backward, as shown, so that it is parallel with the nail taper. Flatten the end of the bar and bend it 90 degrees

THE HOME BLACKSMITH

away from the nail taper to create a stop. Drill a small hole for a screw in the flattened portion, and, when the catch is hammered into a door frame, you can screw-mount the catch so that it will be harder to pull out.

24. Forge a 3-inch dull pointed taper into one end of the latch bar, keeping the thickness the same.

25. Scroll the taper to create a handle to grab from the inside.

26. Tighten up the inside of the scroll so that the tip is pointed inside to prevent anything from becoming hooked on it. Create a double-sided taper to isolate a square of material on the other end to forge round; next, punch and drift a ⅜-inch diameter hole into it.

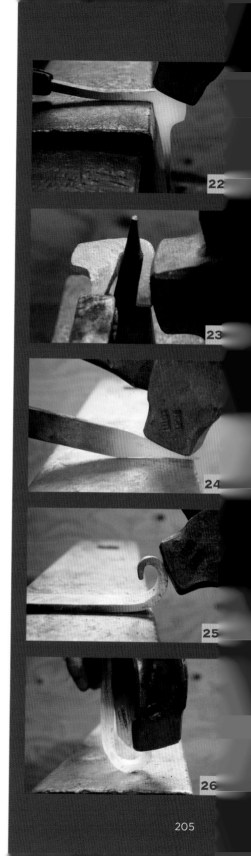

TOILET PAPER HOLDER AND TOWEL BAR SET

STARTING STOCK

- Toilet paper holder: ½-inch square bar, 14 inches long
- Collar on toilet paper holder: ¼ x ½-inch bar, at least 3½ inches long
- Towel bar: ½-inch square bar, 36 inches long

TOOLS REQUIRED

- Hammer
- Anvil
- Softened ball-peen hammer
- ¼-inch ball-nosed punch
- Chisel
- Narrow fuller

1. Create a sharp, square-pointed taper on one end of the toilet paper holder and on both ends of the towel bar.

2. Neck down one side of the ends you tapered, 3 inches from the tip. This photo shows one of the finished pieces. Create a gradual taper into the bar stock further up, about 3 inches.

3. Using your cross-peen, fuller the section you isolated wider. Refine the tip width by turning it on edge as needed. Blend the width into the bar stock about 3 inches further up the bar. Continue to refine the neck until it is a ⅜-inch square right before the flare begins.

   Next, chisel a line along the top edge of the finial and, using a narrow fuller, fuller multiple lines along the bottom edge, perpendicular to the edge. It will curve on its own as the fuller pushes the metal sideways more so than toward the edge. Within 2½ inches of the start of the finial, use a ball-peen hammer to create two evenly spaced divots as you did for the coat hook. Taking a ball-nosed punch, create an even deeper divot to drill out for the mounting screws.

4. The toilet paper holder needs a way to keep the toilet paper on it, so we will forge weld a collar to add the mass to forge into a ball. Begin by tapering one end of the stock for the collar ½ inch long to match one side of the bar that it will be wrapping.

5. Rather than spend time making a collar jig to speed up making one collar, we're just going to bend it around a piece of ½-inch bar as shown. Begin by lining up the tapered edge with one corner of the bar and clamping the two together in the vise for the first bend. After that, clamp the collar in place with one pair of tongs and bend it around with either a hammer or tongs. Before completely closing the last bend, create another ½-inch taper so that it overlaps with the first. Tighten the collar on the anvil and, once it is tight and in place on the end, bring it up to welding heat and weld the collar in place. If it is not tight, you will end up with unwelded gaps. Refine it into a ball shape at welding heat to prevent any weld failures.

6. With the finial decoration facing down, bend the bar 90 degrees over the far edge of the anvil, 1 inch below the last mounting hole. Next, bend it 90 degrees again, 2 inches out, so that it looks like the finished project. Put a slight upward bend to the ball to help keep the roll on the holder. For the towel bar, bend 90 degrees, 3 inches from the last bend on both ends, so that the ends face the same way.

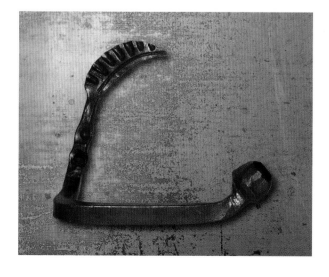

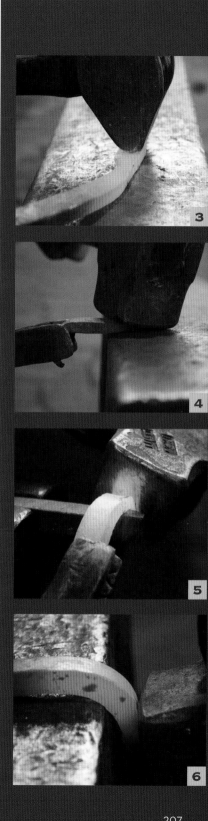

SELLING
YOUR
ITEMS

O nce you start getting better at blacksmithing, it can be a good way to supplement your income. Generally, you'll start selling a few items and taking some orders from friends and family, but, eventually, you can start selling to strangers. Thanks to the Internet, your market can be much larger now than it ever used to be. There are a few tricks to selling your ironwork, however, as you will learn in this chapter.

PHOTOGRAPHY

Photographing your pieces is the most important part of selling your ironwork because you need to showcase all the different items that you can make. Unfortunately, metal is one of the most unforgiving subjects when it comes to photography. It can be difficult to get adequate detail and contrast without harsh shadows or excessive glare.

To get the best photos, you need a sturdy tripod to steady the camera and minimize blurring. This alone will improve the crispness of your photos and make the item "pop" better.

If you are serious about selling your ironwork, look into getting a DSLR camera rather than a simple point-and-shoot camera. I won't go into too much detail here because photography is a craft unto itself, and I can't do it justice in one book, let alone one chapter.

To make a long story short, a DSLR camera gives you more control over what is in focus. DSLR cameras also generally have better sensors, leading to crisper images. That is not to say that buying a better camera will automatically create better photographs; you need to learn a few techniques.

When it comes to photographing ironwork, the two most important considerations are background/composition and lighting. Very often, people photograph their items with too many props or on a busy background, and the actual piece becomes lost in the photo. You want the item you are selling to be the focal point, so use props sparingly, only to draw the viewer's attention to what you are selling.

Lighting is essential to any good photograph and even more so with ironwork. Shadows appear harsher and glare is more obvious when photographing ironwork in direct light. To get the best images, photograph your ironwork in diffuse light, such as sunlight coming through a window with sheer curtains, on a cloudy day, or with a simple light box. There are many plans for easy DIY light boxes; most resemble a tent made of PVC pipe and white muslin or parchment paper that you shine the light through. You need at least two lights with color-matched bulbs, preferably daylight-spectrum bulbs, for a light box to work. To create a seamless background and make photo editing easier, use either poster paper or sheets of cloth hung from the wall and pulled under the item you're photographing.

SELLING LOCALLY

If you want to make it as a blacksmith, you need to sell as often as you can, for as much as you can, so you should maximize the opportunities found in your local area. The first tip is to look into as many local craft fairs and farmers' markets as you can and find out how to exhibit your wares as a vendor. If possible, set up your forge and have someone else handle item sales while you forge. You would be amazed at how many items get stolen if you aren't paying attention, and that severely cuts into your profits. Having your forge going, however, will increase the number of people who visit your table, and they are more likely to buy something that they saw being made.

Once you have a few farmers' markets and craft fairs lined up, you need a sales strategy. I have found that smaller items, in the under-$20 price range,

DID YOU KNOW?

I prefer to use beeswax during my demos to waterproof my forged items during the day—the sweet smell seems to attract clients on its own.

sell the best. Often, blacksmiths get gung-ho about selling bigger pieces that make a larger profit, but most people who are interested in bigger pieces want them to be customized. Your demos and vendor events are great places to advertise your custom sales, but your focus should be on smaller items to sell from your table. You are better off having an album of great photos of your larger pieces rather than having them all on display, only to pack up again at the end of the day.

Another great way to help sell your items is to have storyboards made up for some of them, showing the processes and different steps that go into making the items, especially things that really change the shape of a rod, like a leaf. Odds are that you won't make as much money as you hoped from craft fairs themselves, so treat the vendor fees as advertising expenses. That said, you can sometimes work out a deal with the event's organizers whereby they give you a discount or even free entry if you can set up a demo that will attract attendees to the event.

ONLINE MARKETPLACES

This list is changing constantly as new marketplaces start up, existing online retailers start offering selling options, and other sites merge or shut down, but these are a few of the more common places that I and other blacksmiths use to sell our wares online. There is, of course, eBay, as well as the following sites that cater to handcrafters.

Artfire: www.artfire.com

Big Cartel: www.bigcartel.com

Craft Is Art: www.craftisart.com

DaWanda: en.dawanda.com

Etsy: www.etsy.com

Made it Myself: www.madeitmyself.com

Meylah Marketplaces: www.marketplace.meylah.com

Shophandmade: www.shophandmade.com

Silkfair: www.silkfair.com

Zibbet: www.zibbet.com

ONLINE SHOPS

For many blacksmiths, selling online expands your market greatly. The first step is to set up your own website and a Facebook page for your business; this online presence can help you gain a following of potential customers. It is important to not only sell your work but also give a backstory—people who pay more for handmade items do it for the feeling of getting one-of-a-kind items that were made with a craftsman's care and attention.

Even if you don't have the know-how to set up a free website or the cash to pay someone else to do it for you, there are other ways to easily sell items online. Today, there are many online marketplaces for artisans that let you roll out your virtual carpet and sell for a small fee. Do some research and ask some other craftspeople how they sell online. Eye-catching photographs and engaging descriptions that evoke emotions will not only sell your items but also stand out in the crowd.

ACKNOWLEDGMENTS

This book would not be possible without the help, support, and patience of my loving wife, Lynn, who gave up many weekends and evenings to take the great photographs for this book and wait for me to finish writing. Thank you to my mother and father for letting me get my first anvil and forge and encouraging me while I stumbled through the learning curve of a new blacksmith as well as for encouraging me to write a book on blacksmithing.

I would like to thank my father-in-law, Eric, for the use of his shop while mine was still packed up and in limbo from moving and for his advice throughout the process, as well as my mother-in-law, Gail, for cooking suppers for us after our days in the forge.

Last but not least, I would like to thank my brother-in-law Gary for his help with fetching coal, running the blower, and striking for me.

RESOURCES

This is a list of books and websites I have found very helpful, including some books that are primarily for inspiration rather than education. Many older public-domain books can be found on sites such as Internet Archive and Project Gutenberg, and you may find some that will help keep you motivated and encouraged to advance yourself as a blacksmith.

Books

Aspery, Mark. *The Skills of the Blacksmith Volume I: Mastering the Fundamentals of Blacksmithing*. Self-published, 2007.

——. *The Skills of the Blacksmith Volume II: Mastering the Fundamentals of Leafwork*. Self-published, 2009.

——. *The Skills of the Blacksmith Volume III: Mastering the Fundamentals of Traditional Joinery*. Self-published, 2013.

Bealer, Alex W. *The Art of Blacksmithing*, rev. Castle Books, 2009.

DeLaRonde, Joe. *Blacksmithing Basics for the Homestead*. Castle Books, 2009.

Hrisoulas, Jim. *The Complete Bladesmith*. Boulder, CO: Paladin Press, 1987.

——. *The Master Bladesmith*, Boulder, CO: Paladin Press, 1991.

Rural Development Commission. *The Blacksmith's Craft: An Introduction to Smithing for Apprentices and Craftsmen*. 1997.

Sims, Lorelei. *The Backyard Blacksmith*. Minneapolis: Crestline Books, 2009.

Southworth, Susan and Michael. *Ornamental Ironwork: An Illustrated Guide to its Design, History and Use in American Architecture*. New York: McGraw-Hill, 1992.

Streeter, Donald. *Professional Smithing*. Apple Valley, MN: Astragal Press, 2008.

Weygers, Alexander G. *The Complete Modern Blacksmith*. Berkeley, CA: Ten Speed Press, 1997.

Websites

Anvilfire
www.anvilfire.com

Artist-Blacksmith's Association of North America (ABANA)
www.abana.org

British Artist Blacksmiths Association
www.baba.org.uk

I Forge Iron
www.iforgeiron.com

Internet Archive
www.archive.org

Project Gutenberg
www.gutenberg.org

THE HOME BLACKSMITH

INDEX

Note: Page numbers in **bold** typeface indicate a photograph.

PHOTOS

ABOUT THE AUTHOR

Ryan Ridgway, DVM, grew up on a family farm about an hour south of Regina, Saskatchewan, where he started blacksmithing to make repairs around the farm. During his university years in Saskatoon, he began selling his ironwork online and at local craft fairs to help pay for his degree. During this time, he also took a blacksmithing course offered by his local museum and began volunteering there, which gave him the opportunity to learn from the other blacksmiths and, in turn, teach other new smiths. After receiving his veterinary degree, he began writing blacksmithing articles and has been published in magazines such as *Hobby Farms, Carving Magazine, Canadian Cowboy Country Magazine*, and *Rural Roots.*